McDougal Littell
Pre-Algebra

Larson   Boswell   Kanold   Stiff

CHAPTER 8

Resource Book

The Resource Book contains a wide variety of blackline masters
available for Chapter 8. The blacklines are organized by lesson.
Included are support materials for the teacher as well as practice,
activities, applications, and project resources.

McDougal Littell

A DIVISION OF HOUGHTON MIFFLIN COMPANY

Evanston, Illinois • Boston • Dallas

Contributing Authors

The authors wish to thank **Jessica Pflueger** for her contributions to the Chapter 8 Resource Book.

ISBN-13: 978-0-618-26945-7   ISBN-10: 0-618-26945-2

8 9 10 11 0928 12 11 10 09

Contents

CHAPTER 8 — Linear Functions

Contents

Contents

Descriptions of Resources

This Chapter Resource Book is organized by lessons within the chapter in order to make your planning easier. The following materials are provided:

Tips for New Teachers   These teaching notes provide both new and experienced teachers with useful teaching tips for each lesson, including tips about common errors and inclusion.

Parents as Partners   This guide helps parents contribute to student success by providing an overview of the chapter along with questions and activities for parents and students to work on together.

Lesson Plans and Lesson Plans for Block Scheduling   This planning template helps teachers select the materials they will use to teach each lesson from among the variety of materials available for the lesson. The block-scheduling version provides additional information about pacing.

Activity Masters   These blackline masters make it easier for students to record their work on selected activities in the Student Edition, or they provide alternative activities for selected lessons.

Technology Activities with Keystrokes   Keystrokes for various models of calculators are provided for each Technology Activity in the Student Edition where appropriate, along with alternative Technology Activities for selected lessons.

Practice A, B, and C   These exercises offer additional practice for the material in each lesson, including application problems. There are three levels of practice for each lesson: A (basic), B (average), and C (advanced).

Study Guide   These two pages provide additional instruction, worked-out examples, and practice exercises covering the key concepts and vocabulary in each lesson.

Real-World Problem Solving   These exercises offer problem-solving activities for the material in selected lessons in a real world context.

Challenge Practice   These exercises offer challenging practice on the mathematics of each lesson for advanced students.

Chapter Review Games and Activities   This worksheet offers fun practice at the end of the chapter and provides an alternative way to review the chapter content in preparation for the Chapter Test.

Projects with Rubric   These projects allow students to delve more deeply into a problem that applies the mathematics of the chapter. Teacher's notes and a 4-point rubric are included. The projects include a real-life project, a cooperative project, and an independent extra credit project.

Cumulative Practice   These practice pages help students maintain skills from the current chapter and preceding chapters.

CHAPTER 8 Tips for New Teachers

For use with Chapter 8

Lesson 8.1

COMMON ERROR Students may incorrectly reverse the definition of function. They may think that if two or more inputs give the same output, then the relation is *not* a function. Be certain students understand that Checkpoint Exercise 3 is a function, while Checkpoint Exercise 4 is not.

TEACHING TIP Alphabetically, *domain* precedes *range*, *input* precedes *output*, and *x* precedes *y*. Remembering this can help students keep straight what value goes into a relation and what value comes out.

INCLUSION Visual learners will benefit by graphing ordered pairs in exercises such as Exercises 5 and 6 on page 387. This way they can base their answers on the vertical line test.

Lesson 8.2

COMMON ERROR When students obtain an incorrect value for *y* for a given value of *x*, they may still graph the resulting ordered pair. When it does not form a line with the other ordered pairs, they may still connect the points anyway. Stress that an equation such as $x + 2y = 6$ is a *linear* equation and therefore must graph as a *straight* line. Furthermore, teach students that they should always graph *at least three* ordered pairs when graphing a linear equation. Graphing only two ordered pairs will always create a straight line, but not necessarily the correct one.

TEACHING TIP As a way to recall which lines are horizontal and which are vertical, students should remember the following. Vertical lines only have *x* in their equations, and their graphs only cross the *x*-axis. Horizontal lines only have *y* in their equations, and their graphs only cross the *y*-axis.

TEACHING TIP Stress the definition of function form in the middle of page 393 as this will lead into slope-intercept form in Lesson 8.5 and function notation in Lesson 8.7.

TEACHING TIP When the independent variable has a fractional coefficient, you may wish to encourage students to choose values for the variable that are multiples of the denominator of the fraction. For instance, in Example 5 the equation is $y = \frac{1}{2}x + 3$, so students could choose $-4, -2, 0, 2$, and 4 for *x* in order to graph integers rather than fractions or mixed numbers.

Lesson 8.3

TEACHING TIP In purely mathematical situations such as Example 2 on page 399, stress that a line is an infinite object. Students should graph the line not only in Quadrant IV, but also should extend the line into Quadrants I and II as well. However, in real-life situations, such as Example 3, the graph should be drawn only in Quadrant I as the two quantities, *paddling time* and *drifting time*, can not be negative.

COMMON ERROR Students often confuse which ordered pairs match which intercept. Stress that *x*-intercepts lie on the *x*-axis and, therefore, have *y*-coordinates of 0. Likewise, *y*-intercepts lie on the *y*-axis and therefore have *x*-coordinates of 0.

Lesson 8.4

COMMON ERROR Some students may incorrectly use the reciprocal of the slope formula. Ask those students if they can *run* while they are seated. When they respond in the negative, point out that they must first *rise* before they can *run*. Since *rise* must come first, it is written first, in the numerator.

COMMON ERROR Be certain students understand that if they take y_2 from one particular ordered pair, then they must take x_2 from the same ordered pair. Sometimes students will switch the order of subtraction to make computation easier. Show them why they cannot do this.

Tips for New Teachers

For use with Chapter 8

TEACHING TIP To help students remember that horizontal lines have a slope of 0 and that vertical lines have an undefined slope, try the following scenario. Have students imagine that these lines are ski slopes. If they tried to ski a horizontal slope, they would go nowhere, or 0. If they tried to ski a vertical slope, they would fall off and end up undefined.

Lesson 8.5

TEACHING TIP In Checkpoint Exercise 3 on page 413, be certain students understand that a slope of 4 should be interpreted as $\frac{4}{1}$ for the purposes of graphing. More importantly, be certain students understand that the slope in Checkpoint Exercise 1 is -1 which should be interpreted as $\frac{-1}{1}$, or $\frac{1}{-1}$.

COMMON ERROR Students may be confused by the lack of a b term in Checkpoint Exercise 3. Be certain students understand that the b term, or y-intercept, is 0.

COMMON ERROR When finding slopes of perpendicular lines, students often forget one of the two parts of negative reciprocal. You may wish to break this concept down into two steps. First ask students what the reciprocal of the slope would be and then ask them what the negative of that reciprocal would be.

Lesson 8.6

TEACHING TIP In Example 4 on page 420 students should note that the slope of a line does not depend on which two points are used in the slope formula. Elaborate upon this as necessary. Make certain students understand that the slope is the same for the infinite length of the line. Ask students to choose three or four different points from the graph and have them calculate the slope between different pairs of these points to see that the slope is always the same.

Lesson 8.7

COMMON ERROR Students may confuse function notation $f(x)$ with multiplication. Be certain students understand the notation before continuing the lesson.

Lesson 8.8

COMMON ERROR When checking a solution to a system of equations, students may only check their answer in one of the two equations and stop if the answer checks. Make certain students check the point of intersection in both equations.

Lesson 8.9

TEACHING TIP The ordered pair (0, 0) is the easiest point to use in determining which side of a line to shade in a linear inequality. This point can be used in all cases, except when the line intersects the origin.

Name _____ Date _____

Parents as Partners
For use with Chapter 8

Chapter Overview One way you can help your student succeed in Chapter 8 is by discussing the lesson goals in the chart below. When a lesson is completed, ask your student the following questions. "What were the goals of the lesson? What new words and formulas did you learn? How can you apply the ideas of the lesson to your life?"

Lesson Title	Lesson Goals	Key Applications
8.1: Relations and Functions	Use graphs to represent relations and functions.	• Alligators • Hurricanes • Skydivers • Birds
8.2: Linear Equations in Two Variables	Find solutions of equations in two variables.	• Volcanoes • Spacecraft • Platypuses
8.3: Using Intercepts	Use x- and y-intercepts to graph linear equations.	• Shopping • Animal Nutrition • Transportation • Fitness
8.4: The Slope of a Line	Find and interpret slopes of lines.	• Wakeboarding • Animal Speeds • Cinder Cones • Roads
8.5: Slope-Intercept Form	Graph linear equations in slope-intercept form.	• Knitting • Paramotoring • Walk-a-thon
8.6: Writing Linear Equations	Write linear equations.	• Bamboo • Clams • Glaciers
8.7: Function Notation	Use function notation.	• Cable TV • Squid • Rivers
8.8: Systems of Linear Equations	Graph and solve systems of linear equations.	• Shoes • Vacation Rentals • Inkjet Printers
8.9: Graphs of Linear Inequalities	Graph inequalities in two variables.	• Pottery • Entertainment • Kites

Notetaking Strategies

Using Color is the strategy featured in Chapter 8 (see page 384). Encourage your student to use color to identify important pieces of information and show how they are related to each other in his/her notes. Your student can use color when finding slopes of lines in Lesson 8.4.

Name _____ Date _____

CHAPTER 8 | Parents as Partners

For use with Chapter 8

Key Ideas Your student can demonstrate understanding of key concepts by working through the following exercises with you.

Lesson	Exercise
8.1	Identify the domain and range of the relation. Then tell whether the relation is a function. (a) $(3, 5), (-5, 3), (2, 5), (3, -4), (5, 5)$ (b) $(-2, 0), (0, 1), (2, 2), (4, 3), (6, 4)$
8.2	Tell whether the ordered pair is a solution of the equation. (a) $y = 5x - 7$; $(3, 8)$ (b) $2x + 4y = -16$; $(-9, 1)$
8.3	Find the intercepts of the graph of $-6x + 3y = 12$.
8.4	Find the slope of the line through the given points. Then tell whether the slope is *positive*, *negative*, *zero*, or *undefined*. (a) $(2, 5), (6, 1)$ (b) $(-1, 4), (3, 4)$ (c) $(0, 2), (5, 12)$ (d) $(-3, -1), (-3, 6)$
8.5	Identify the slope and y-intercept of the line with the given equation. (a) $y = 3x + 2$ (b) $9x - 6y = 18$
8.6	Write an equation of the line through the given points. (a) $(0, -4), (3, -2)$ (b) $(-2, 12), (0, 2)$
8.7	Let $f(x) = 2x + 8$. Find the indicated value. (a) $f(x)$ when $x = 7$ (b) $f(-9)$ (c) x when $f(x) = 44$
8.8	Tell whether $(8, 12)$ is a solution of $-3x + 5y = 36$ and $2x - 2y = 8$.
8.9	Tell whether the ordered pair is a solution of the inequality. (a) $y \leq -4x + 4$; $(2, -2)$ (b) $2y - 8x > -45$; $(4, 6)$

Home Involvement Activity

Directions: Collect data on the age and height of people in your household and neighborhood. Make a chart that contains the data you gathered. Represent this data as a graph and as a mapping diagram. Is your relation a function? Explain your reasoning.

Answers: 8.1: (a) Domain: $-5, 2, 3, 5$; Range: $-4, 3, 5$; not a function (b) Domain: $-2, 0, 2, 4, 6$; Range: $0, 1, 2, 3, 4$; function **8.2:** (a) solution (b) not a solution **8.3:** x-intercept: -2; y-intercept: -2; **8.4:** (a) $m = -1$; negative (b) $m = 0$; zero (c) $m = 2$; positive (d) undefined **8.5:** (a) $m = 3$; y-intercept: 2 (b) $m = \frac{3}{2}$; y-intercept: -3 **8.6:** (a) $y = \frac{2}{3}x - 4$ (b) $y = -5x + 2$ **8.7:** (a) 22 (b) -10 (c) 18 **8.8:** not a solution **8.9:** (a) not a solution (b) solution

Teacher's Name _____   Class _____   Room _____   Date _____

Lesson Plan

1-day lesson (See *Pacing and Assignment Guide*, TE page 382A)

For use with pages 385–390

GOAL   **Use graphs to represent relations and functions.**

State/Local Objectives _____

_____

✓ **Check the items you wish to use for this lesson.**

STARTING OPTIONS

____ Warm-Up: Transparencies

TEACHING OPTIONS

____ Notetaking Guide

____ Examples: 1–4, SE pages 385–387

____ Extra Examples: TE pages 386–387

____ Checkpoint Exercises: 1–4, SE pages 385–386

____ Concept Check: TE page 387

____ Guided Practice Exercises: 1–7, SE page 387

APPLY/HOMEWORK

Homework Assignment

____ Basic: EP p. 803 Exs. 38–41; pp. 388–390 Exs. 8–16, 18–21, 27–37

____ Average: pp. 388–390 Exs. 10–17, 20–24, 26–37

____ Advanced: pp. 388–390 Exs. 10–18, 21–28*, 33–37

Reteaching the Lesson

____ Practice: CRB pages 7–9 (Level A, Level B, Level C); Practice Workbook

____ Study Guide: CRB pages 10–11; Spanish Study Guide

Extending the Lesson

____ Challenge: SE page 390; CRB page 12

ASSESSMENT OPTIONS

____ Daily Quiz (8.1): TE page 390 or Transparencies

____ Standardized Test Practice: SE page 390

Notes

_____

_____

_____

_____

Teacher's Name _____ Class _____ Room _____ Date _____

Lesson Plan for Block Scheduling

Half-block lesson (See *Pacing and Assignment Guide*, TE page 382A)

For use with pages 385–390

GOAL Use graphs to represent relations and functions.

State/Local Objectives _____

_____

✓ **Check the items you wish to use for this lesson.**

STARTING OPTIONS

_____ Warm-Up: Transparencies

TEACHING OPTIONS

_____ Notetaking Guide

_____ Examples: 1–4, SE pages 385–387

_____ Extra Examples: TE pages 386–387

_____ Checkpoint Exercises: 1–4, SE pages 385–386

_____ Concept Check: TE page 387

_____ Guided Practice Exercises: 1–7, SE page 387

Chapter Pacing Guide	
Day	Lesson
1	**8.1**; 8.2 (begin)
2	8.2 (end); 8.3
3	8.4; 8.5
4	8.6
5	8.7; 8.8
6	8.9
7	Ch. 8 Review and Projects

APPLY/HOMEWORK

Homework Assignment

_____ Block Schedule: pp. 388–390 Exs. 10–17, 20–24, 26–37 (with 8.2)

Reteaching the Lesson

_____ Practice: CRB pages 7–9 (Level A, Level B, Level C); Practice Workbook

_____ Study Guide: CRB pages 10–11; Spanish Study Guide

Extending the Lesson

_____ Challenge: SE page 390; CRB page 12

ASSESSMENT OPTIONS

_____ Daily Quiz (8.1): TE page 390 or Transparencies

_____ Standardized Test Practice: SE page 390

Notes

_____

_____

_____

_____

_____

_____

Practice A

For use with pages 385–390

Complete the statement.

1. Each number in the __?__ of a relation is an input.

2. Each number in the __?__ of a relation is an output.

Identify the domain and range of the relation.

3. (0, 2), (1, 4), (2, 6), (3, 8), (4, 10)

4. (3, 4), (3, 5), (4, 6), (4, 7), (5, 8)

5.

x	0	2	2	3
y	5	5	6	6

6.

x	1	4	7	10	13
y	−5	−4	−3	−2	−1

Represent the relation as a graph and as a mapping diagram. Then tell whether the relation is a function. Explain your reasoning.

7. (−2, 1), (0, 4), (1, 3), (2, 3)

8. (2, −1), (2, 1), (3, 2), (3, 0), (4, 0)

9. (−3, 0), (−2, 0), (−1, 1), (0, 1)

10. (0, 1), (0, 2), (1, 3), (1, 4), (1, 5)

11.

x	−2	−1	0	1
y	−1	2	4	5

12.

x	5	6	6	8	8
y	3	3	4	4	5

13.

x	4	6	8	10
y	4	6	4	6

14.

x	−1	−1	0	2	3
y	3	4	3	4	5

In Exercises 15–17, tell whether the relation represented by the graph is a function.

15.
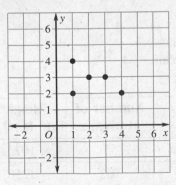

16.
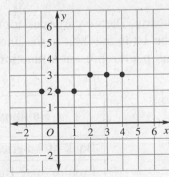

17.
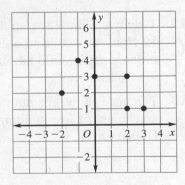

18. The amount of precipitation in millimeters is measured every day from day 1 to day 20. Do the ordered pairs (day, precipitation) represent a function? Explain.

19. Terrance receives $.50 per pound of aluminum cans he recycles. Is the number of cans he recycles a function of the amount of money he receives? Explain.

Name _____ Date _____

Practice B

For use with pages 385–390

Identify the domain and range of the relation.

1. (5, 0), (6, 0), (7, 6), (8, 8), (8, 10)

2. (−6, 4), (−3, 0), (4, 2), (4, 3), (7, 9)

3.

x	−4	−4	−1	3	4
y	−5	−4	−3	2	0

4.

x	0	0	2	4	8
y	−3	−1	1	3	−1

Represent the relation as a graph and as a mapping diagram. Then tell whether the relation is a function. Explain your reasoning.

5. (−2, 2), (−1, 2), (1, 2), (2, 2)

6. (0, 0), (1, 1), (1, 2), (3, 3), (4, 4)

7.

x	−2	−1	0	1	2
y	1	2	2	1	0

8.

x	2	4	1	2	5
y	3	1	3	2	4

In Exercises 9–11, tell whether the relation represented by the graph is a function.

9.

10.

11.

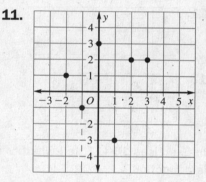

12. Twenty children line up to ride go-carts. The go-cart operator collects $2 from each child in order from the 1st to the 20th in the line. Do the ordered pairs (child number, amount paid) represent a function? Explain your reasoning.

13. The table shows the number of stories and height of five buildings in the United States.

Building	Number of stories, x	Height (in feet), y
Bank of America Plaza	55	1023
Empire State Building	102	1250
Library Tower	75	1018
Sears Tower	110	1450
JP Morgan Chase Tower	75	1002

a. Identify the domain and range of the relation given by the ordered pairs (x, y).

b. Draw a mapping diagram for the relation.

c. Is the relation a function? Explain.

Practice C

For use with pages 385–390

Identify the domain and range of the relation.

1. (5, 0), (1, 4), (7, 0), (2, 4), (4, 0)

2. (5, −1), (−2, 7), (−9, −2), (7, 6), (2, −2)

3.

x	−1	8	−6	4	1
y	−1	−4	6	−4	2

4.

x	−3	3	−2	2	0
y	−9	−1	3	−3	6

Represent the relation as a graph and as a mapping diagram. Then tell whether the relation is a function. Explain your reasoning.

5. (5, 8), (6, 7), (7, 7), (8, 8)

6. (4, 2), (−1, 4), (1, 3), (2, 3), (4, 1)

Represent the relation shown in the graph as a set of ordered pairs. Then tell if the relation is a function. Explain your reasoning.

7.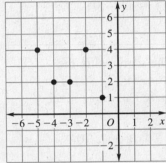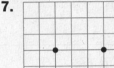

8.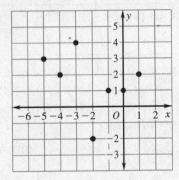

9.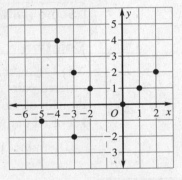

10. You make $1 per quart picking strawberries. Do the ordered pairs (number of quarts, earnings) represent a function? Explain your reasoning.

11. Your aunt works at a factory when they need extra help. She makes $8 per hour on first shift, $9 per hour on second shift, and $10 per hour on third shift.

   a. Do the ordered pairs (hours worked, earnings) represent a function for all possible values? Explain.

   b. Do the ordered pairs (earnings, hours worked) represent a function for all possible values? Explain.

   c. Explain a way you could divide the ordered pairs in part (a) into three sets that are functions.

12. Explain why the distance you travel on your bike is not always a function of the time you spend riding your bike. Describe a situation when it would be a function.

Name _____   Date _____

Study Guide

For use with pages 385–390

GOAL Use graphs to represent relations and functions.

> ### VOCABULARY
>
> A **relation** is a pairing of numbers in one set, called the **domain,** with numbers in another set, called the **range.** Each number in the domain is an **input.** Each number in the range is an **output.**

EXAMPLE 1 Identifying the Domain and Range

The table below shows the scores, with respect to par, of a player's first 7 golf matches of the season. Identify the domain and range of the relation.

Match number, x	1	2	3	4	5	6	7
Score (with respect to par), y	10	7	5	−3	0	−3	−5

Solution

You can represent the relationship between the match and the score using the ordered pairs (x, y): (1, 10), (2, 7), (3, 5), (4, −3), (5, 0), (6, −3), (7, −5). The domain of the relation is the set of all inputs, or x-coordinates. The range is the set of all outputs, or y-coordinates.

  Domain: 1, 2, 3, 4, 5, 6, 7   **Range:** −5, −3, 0, 5, 7, 10

EXAMPLE 2 Representing a Relation

Represent the relation (−7, −2), (−6, 0), (−5, −1), (−5, −3), (−2, 1), (1, 1) as indicated.

  a. A graph   **b.** A mapping diagram

Solution

a. Graph the ordered pairs as points in a coordinate plane.

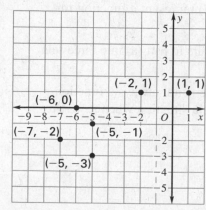

b. List the inputs and the outputs in numerical order. Draw arrows from the inputs to their outputs.

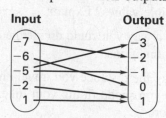

Name _____ Date _____

Study Guide
For use with pages 385–390

Exercises for Examples 1 and 2

Identify the domain and range of the relation. Then represent the relation as a graph and as a mapping diagram.

1. (−1, 0), (0, 8), (2, −3), (3, 4), (3, 7)   **2.** (−4, −6), (−1, −2), (5, −8), (5, −2)

VOCABULARY

A relation is a **function** if for each input there is *exactly one* output.

When a relation is represented by a graph, you can use the *vertical line test* to tell whether the relation is a function. The **vertical line test** says that if you can find a vertical line passing through more than one point of the graph, then the relation *is not* a function. Otherwise, the relation *is* a function.

EXAMPLE 3 **Identifying Functions**

Tell whether the relation is a function.

a. The relation in Example 1, consisting of the ordered pairs (match, score) for 7 matches:

(1, 10), (2, 7), (3, 5), (4, −3), (5, 0), (6, −3), (7, −5)

b. The relation in Example 2, consisting of the following ordered pairs:

(−7, −2), (−6, 0), (−5, −1), (−5, −3), (−2, 1), (1, 1)

Solution

a. The relation *is* a function because every input is paired with exactly one output. This makes sense, because you can only have one score per match.

b. The relation *is not* a function because the input −5 is paired with two outputs, −1 and −3.

Exercises for Example 3

Tell whether the relation is a function. Explain your reasoning.

3. (−10, 5), (2, 2), (−1, 0), (0, −1), (−10, 4)

4. (8, 3), (−1, 3), (2, 4), (6, 3)

LESSON
8.1 Challenge Practice

Name _____ Date _____

For use with pages 385–390

Identify the domain and range of the relation. Then tell whether the relation is a function.

1.

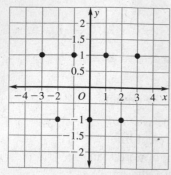

2.

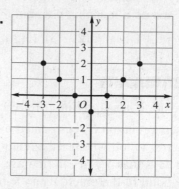

3.

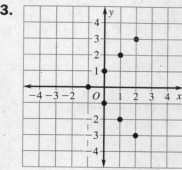

Represent the relation as a graph and as a mapping diagram. Then tell whether the relation is a function. Explain your reasoning.

4. $(-16, 8), (-8, 16), (16, 4), (-4, 2), (2, -4), (4, -2)$

5. $(-4, 6.25), (3, -2.5), (0, -0.75), (-3, -2.5), (4, 6.25)$

In Exercises 6–8, determine whether the relation described could be a function. Explain your reasoning.

6. The amount of rainfall in your town is recorded every day for a month. A relation is given by the ordered pairs (inches of rain, day of month).

7. You record the number of pages you write in your journal each day for a month. A relation is given by the ordered pairs (day of month, pages).

8. You record how many miles you run each day and the amount of time it takes you to run that distance. A relation is given by the ordered pairs (distance, time).

Teacher's Name _____ Class _____ Room _____ Date _____

Lesson Plan

2-day lesson (See *Pacing and Assignment Guide*, TE page 382A)

For use with pages 391–397

GOAL   **Find solutions of equations in two variables.**

State/Local Objectives _____

_____

✓ **Check the items you wish to use for this lesson.**

STARTING OPTIONS

____ Homework Check (8.1): TE page 388; Answer Transparencies

____ Homework Quiz (8.1): TE page 390; Transparencies

____ Warm-Up: Transparencies

TEACHING OPTIONS

____ Notetaking Guide

____ Examples: Day 1: 1–2, SE pages 391–392; Day 2: 3–5, SE pages 392–393

____ Extra Examples: TE pages 392–393

____ Checkpoint Exercises: Day 1: 1–4, SE page 391; Day 2: 5–10, SE pages 392–393

____ Concept Check: TE page 393

____ Guided Practice Exercises: Day 1: 3–6, SE page 394; Day 2: 1–2, 7–11, SE page 394

____ Technology Activity: SE page 397

APPLY/HOMEWORK

Homework Assignment

____ Basic: Day 1: SRH p. 780 Exs. p. 9–12; pp. 394–396 Exs. 12–15, 32–38, 43–50
Day 2: pp. 394–396 Exs. 16–27, 51–56

____ Average: Day 1: pp. 394–396 Exs. 12–15, 32–38, 43–50
Day 2: pp. 394–396 Exs. 20–31, 39–41, 51–56

____ Advanced: Day 1: pp. 394–396 Exs. 12–15, 32–34, 37, 38, 45–48
Day 2: pp. 394–396 Exs. 20–27, 39–44*, 53–56

Reteaching the Lesson

____ Practice: CRB pages 15–17 (Level A, Level B, Level C); Practice Workbook

____ Study Guide: CRB pages 18–19; Spanish Study Guide

Extending the Lesson

____ Challenge: SE page 396; CRB page 20

ASSESSMENT OPTIONS

____ Daily Quiz (8.2): TE page 396 or Transparencies

____ Standardized Test Practice: SE page 396

Notes _____

Teacher's Name _____ Class _____ Room _____ Date _____

Lesson Plan for Block Scheduling

1-block lesson (See *Pacing and Assignment Guide*, TE page 382A)

For use with pages 391–397

GOAL   Find solutions of equations in two variables.

State/Local Objectives _____

_____

✓ Check the items you wish to use for this lesson.

Chapter Pacing Guide	
Day	**Lesson**
1	8.1; **8.2 (begin)**
2	**8.2 (end)**; 8.3
3	8.4; 8.5
4	8.6
5	8.7; 8.8
6	8.9
7	Ch. 8 Review and Projects

STARTING OPTIONS

____ Homework Check (8.1): TE page 388; Answer Transparencies

____ Homework Quiz (8.1): TE page 390; Transparencies

____ Warm-Up: Transparencies

TEACHING OPTIONS

____ Notetaking Guide

____ Examples: Day 1: 1–2, SE pages 391–392;
Day 2: 3–5, SE pages 392–393

____ Extra Examples: TE pages 392–393

____ Checkpoint Exercises: Day 1: 1–4, SE page 391;
Day 2: 5–10, SE pages 392–393

____ Concept Check: TE page 393

____ Guided Practice Exercises: Day 1: 3–6, SE page 394;
Day 2: 1–2, 7–11, SE page 394

____ Technology Activity: SE page 397

APPLY/HOMEWORK

Homework Assignment

____ Block Schedule: Day 1: pp. 394–396 Exs. 12–15, 32–38, 43–50 (with 8.1)
Day 2: pp. 394–396 Exs. 20–31, 39–41, 51–56 (with 8.3)

Reteaching the Lesson

____ Practice: CRB pages 15–17 (Level A, Level B, Level C); Practice Workbook

____ Study Guide: CRB pages 18–19; Spanish Study Guide

Extending the Lesson

____ Challenge: SE page 396; CRB page 20

ASSESSMENT OPTIONS

____ Daily Quiz (8.2): TE page 396 or Transparencies

____ Standardized Test Practice: SE page 396

Notes

_____

Name _____ Date _____

Practice A

For use with pages 391–397

1. Complete the statement: An equation that is solved for y is in ___?___ form.

Tell whether the ordered pair is a solution of $y = -3x + 9$.

2. $(2, 3)$　　　　　**3.** $(3, 1)$　　　　　**4.** $(1, 3)$　　　　　**5.** $(-3, 18)$

Tell whether the ordered pair is a solution of the equation.

6. $y = x; (-3, 3)$　　　　　　　　　　**7.** $y = 3x + 7; (-2, 1)$

8. $2x - y = 5; (4, 3)$　　　　　　　　**9.** $3y + x = 18; (4, 4)$

Find the value of y when x has the given value in the equation.

10. $y = 3x + 8; x = 4$　　　　　　　　**11.** $y = 4x - 12; x = 6$

12. $y = 22 - 5x; x = 3$　　　　　　　**13.** $y = 8 + 6x; x = -3$

Graph the equation.

14. $y = x + 4$　　　　　**15.** $y = -x + 2$　　　　　**16.** $y = 2x - 1$

17. $y = \frac{1}{2}x + 4$　　　**18.** $y = 3$　　　　　　　**19.** $y = -2$

20. $x = -5$　　　　　　**21.** $x = 10$　　　　　　**22.** $y = 2x$

Write the equation in function form. Then graph the equation.

23. $x + y = 8$　　　　　**24.** $5x - y = -1$　　　　**25.** $y - 2x = 6$

26. $4x + 4y = 8$　　　　**27.** $2y + 3x = 2$　　　　**28.** $3x - 3y = 7$

29. The formula $y = 0.001x$ converts a distance x in meters to a distance y in kilometers. The total distance of eighteen holes at one golf course is 6460 meters. What is this distance in kilometers?

30. Your family is driving on the highway. The number of miles y you travel in x minutes is approximated by the equation $y = 1.1x$. Approximately how far do you travel in 80 minutes?

LESSON
8.2 **Practice B**

For use with pages 391–397

Name _____ Date _____

Tell whether the ordered pair is a solution of the equation.

1. $y = 5x; (15, -3)$

2. $y = 4x + 9; (-2, 1)$

3. $4x - 5y = 1; (4, 3)$

4. $7y - 3x = 11; (5, 8)$

Find the value of d when r has the given value in the equation.

5. $d = 2.5r; r = 64$

6. $d = 3r + 120; r = 62$

7. $d - 5r = 40; r = 4$

8. $12r - d = -240; r = 9$

Graph the equation. Tell whether the equation is a function.

9. $y = x - 3$

10. $y = 2x + 4$

11. $y = -\frac{3}{4}x$

12. $y = -\frac{1}{3}x + 2$

13. $x = -11$

14. $y = 8$

15. $x = 8$

16. $y = -1$

17. $y = 2(x + 1)$

Write the equation in function form. Then graph the equation.

18. $7x - y = 0$

19. $15x + y = 20$

20. $y + 6x - 12 = 0$

21. $6y - 3x = 12$

22. $3x - 2y = 6$

23. $4x - 12y + 24 = 0$

24. The formula $y = 2.205x$ converts a mass x in kilograms to a weight y in pounds. A sports car has a mass of 1270 kilograms. What is its weight in pounds?

25. A high school booster club sets up an academic scholarship that is awarded to one student each year. The formula $y = 2700x$ can be used to find the total amount y of money awarded through this scholarship after x years. What is the total amount of scholarship money paid after 12 years?

Find the value of a that makes the ordered pair a solution of the equation.

26. $y = 3x + 7; (-3, a)$

27. $y = 11 - 7x; (a, -10)$

28. $2x + 4y = 14; (-5, a)$

29. $9x - 5y = -9; (a - 1, 9)$

Lesson 8.2

16 **Pre-Algebra**
Chapter 8 Resource Book

Practice C

For use with pages 391–397

Tell whether the ordered pair is a solution of the equation.

1. $y = 5x + 16$; $(3, 31)$

2. $y = -9x - 12$; $(-3, -39)$

3. $6x - 8y = 22$; $(-7, -8)$

4. $11y - 3x + 17 = 0$; $(20, 7)$

Find the value of d when r has the given value in the equation.

5. $d = \dfrac{3}{5}r$; $r = -55$

6. $d = 7r - 57$; $r = 13$

7. $-24r - 4d = 80$; $r = -7$

8. $d - \dfrac{7}{8}r = 23$; $r = 48$

Graph the equation. Tell whether the equation is a function.

9. $y = 9x$

10. $y = -\dfrac{3}{8}x$

11. $y = -\dfrac{1}{6}x - 2$

12. $y = \dfrac{8}{5}x + 5$

13. $y = \dfrac{4}{3}(3x - 6)$

14. $x = -4.8$

15. $x = \dfrac{7}{2}$

16. $y = -5.5$

17. $y = -\dfrac{13}{3}$

Write the equation in function form. Then graph the equation.

18. $x + y = 1$

19. $x - y = 6$

20. $7x - 5y = 10$

21. $11x + 4y = -12$

22. $-8y + 9x - 4 = 0$

23. $4(x + 3y) - 24 = 0$

24. How could you use a coordinate grid to show that an equation in two variables x and y is not a linear equation?

25. The volume of water y (in cubic meters) that flows over Niagara Falls during peak daytime tourist hours in x minutes can be approximated by the equation $y = 168,000x$. Approximately how many cubic meters of water flow over the falls in one hour?

26. One metric ton is equivalent to about 2204.6 pounds.

   a. Write a linear equation that can be used to convert a mass x in metric tons to a weight y in pounds.

   b. Find the number of pounds equivalent to 0.6 metric ton.

Lesson 8.2

Name _____ Date _____

Study Guide
For use with pages 391–397

GOAL Find solutions of equations in two variables.

> **VOCABULARY**
>
> An example of an **equation in two variables** is $2x - y = 5$. A **solution** of an equation in x and y is an ordered pair (x, y) that produces a true statement when the values of x and y are substituted into the equation.
>
> The **graph** of an equation in two variables is the set of points in a coordinate plane that represent all the solutions of the equation. An equation whose graph is a line is called a **linear equation**.
>
> An equation that is solved for y is in **function form**.

EXAMPLE 1 Checking a Solution

Tell whether $(-1, 19)$ is a solution of $5x + 6y = 100$.

$$5x + 6y = 100 \qquad \text{Write original equation.}$$
$$5(-1) + 6(19) \stackrel{?}{=} 100 \qquad \text{Substitute } -1 \text{ for } x \text{ and } 19 \text{ for } y.$$
$$109 \neq 100 \qquad \text{Simplify.}$$

Answer: $(-1, 19)$ is not a solution of $5x + 6y = 100$.

EXAMPLE 2 Finding Solutions

You are saving money for a stereo system. So far, you have $40 saved. For every lawn you mow, you earn $20. Use the equation $t = 40 + 20m$, where t represents the total amount saved and m is the number of lawns mowed.

a. Make a table of solutions for the equation.

b. How many lawns will you have to mow to save a total of $400?

Solution

a. Substitute values of m into the equation $t = 40 + 20m$, and find values of t. The table shows that the following ordered pairs are solutions of the equation: $(0, 40)$, $(10, 240)$, $(25, 540)$

m	Substitution	t
0	$t = 40 + 20(0)$	40
10	$t = 40 + 20(10)$	240
25	$t = 40 + 20(25)$	540

b. Find the value of m when $t = 400$.

$$400 = 40 + 20m \qquad \text{Substitute 400 for } t \text{ in the equation } t = 40 + 20m.$$
$$360 = 20m \qquad \text{Subtract 40 from each side.}$$
$$18 = m \qquad \text{Divide each side by 20.}$$

Answer: You will have to mow 18 lawns to save a total of $400.

Name _____ Date _____

Study Guide

For use with pages 391–397

Exercises for Examples 1 and 2

Tell whether the ordered pair is a solution of $2y - 9x = -11$.

1. $(5, 17)$ **2.** $(26, 7)$ **3.** $(-1, 10)$ **4.** $(3, 8)$

Make a table of solutions for the equation. Then find x when $y = 0$.

5. $3x + y = 15$ **6.** $4y - 5x = 16$

EXAMPLE 3 **Graphing a Linear Equation**

Write $4y - 12x = 8$ in function form. Then graph the equation.

Solution

To write the equation in function form, solve for y.

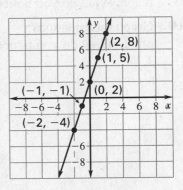

$4y - 12x = 8$ Write original equation.

$4y = 12x + 8$ Add $12x$ to each side.

$y = 3x + 2$ Divide each side by 4.

To graph the equation, use its function form to make a table of solutions. Graph the ordered pairs (x, y) from the table, and draw a line through the points.

x	-2	-1	0	1	2
y	-4	-1	2	5	8

EXAMPLE 4 **Graphing Horizontal and Vertical Lines**

Graph $x = -1$ and $y = 5$.

a. The graph of the equation $x = -1$ is the vertical line through $(-1, 0)$.

b. The graph of the equation $y = 5$ is the horizontal line through $(0, 5)$.

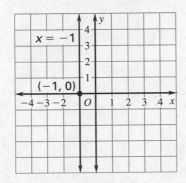
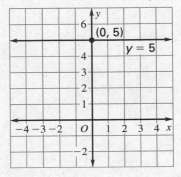

Exercises for Examples 3 and 4

Graph the equation.

7. $5x + 2y = 10$ **8.** $2y - 7x = 6$

9. $x = 3$ **10.** $y = -7$

Challenge Practice

For use with pages 391–397

Write the equation in function form. Then graph the equation.

1. $\frac{2}{3}x + \frac{1}{2}y = -4$

2. $0.2y - 4x = -6.8$

3. $-\frac{3}{8}x - \frac{2}{5}y = \frac{1}{2}$

4. $1.23 - 3y = 2.1x$

Find the value of a that makes the ordered pair a solution of the equation.

5. $5x - 2y = 8;\ (-2, a)$

6. $6y - 11x = -4;\ (a - 1, -8)$

7. $-4x - 3y = 17;\ (a + 1, 5)$

8. $2.1x - 3.4y = 3.2;\ (a + 3, 4)$

9. Graph the equations $y = x + 2$ and $x + y = 3$ in the same coordinate plane. What do you notice about the graphs?

10. Graph the equations $y = x + 2$ and $y + 1 = x$ in the same coordinate plane. What do you notice about the graphs?

Lesson 8.2

Teacher's Name _____   Class _____   Room _____   Date _____

Lesson Plan

1-day lesson (See *Pacing and Assignment Guide*, TE page 382A)

For use with pages 398–402

GOAL   Use *x*- and *y*-intercepts to graph linear equations.

State/Local Objectives   _____

_____

✓ Check the items you wish to use for this lesson.

STARTING OPTIONS

_____ Homework Check (8.2): TE page 394; Answer Transparencies

_____ Homework Quiz (8.2): TE page 396; Transparencies

_____ Warm-Up: Transparencies

TEACHING OPTIONS

_____ Notetaking Guide

_____ Examples: 1–3, SE pages 398–399

_____ Extra Examples: TE page 399

_____ Checkpoint Exercises: 1–3, SE page 399

_____ Concept Check: TE page 399

_____ Guided Practice Exercises: 1–9, SE page 400

APPLY/HOMEWORK

Homework Assignment

_____ Basic: SRH p. 785 Exs. 1, 4, 5, 7; pp. 400–402 Exs. 10–19, 22–25, 30, 34–47

_____ Average: pp. 400–402 Exs. 13–24, 27–32, 36–39, 42–48

_____ Advanced: pp. 400–402 Exs. 15–20, 23–35*, 40–48

Reteaching the Lesson

_____ Practice: CRB pages 23–25 (Level A, Level B, Level C); Practice Workbook

_____ Study Guide: CRB pages 26–27; Spanish Study Guide

Extending the Lesson

_____ Challenge: SE page 402; CRB page 28

ASSESSMENT OPTIONS

_____ Daily Quiz (8.3): TE page 402 or Transparencies

_____ Standardized Test Practice: SE page 402

Notes _____

_____

_____

_____

Teacher's Name _____ Class _____ Room _____ Date _____

Lesson Plan for Block Scheduling

Half-block lesson (See *Pacing and Assignment Guide*, TE page 382A)

For use with pages 398–402

GOAL   Use *x*- and *y*-intercepts to graph linear equations.

State/Local Objectives _____

_____

✓ **Check the items you wish to use for this lesson.**

STARTING OPTIONS

_____ Homework Check (8.2): TE page 394; Answer Transparencies

_____ Homework Quiz (8.2): TE page 396; Transparencies

_____ Warm-Up: Transparencies

TEACHING OPTIONS

_____ Notetaking Guide

_____ Examples: 1–3, SE pages 398–399

_____ Extra Examples: TE page 399

_____ Checkpoint Exercises: 1–3, SE page 399

_____ Concept Check: TE page 399

_____ Guided Practice Exercises: 1–9, SE page 400

APPLY/HOMEWORK

Homework Assignment

_____ Block Schedule: pp. 400–402 Exs. 13–24, 27–32, 36–48 (with 8.2)

Reteaching the Lesson

_____ Practice: CRB pages 23–25 (Level A, Level B, Level C); Practice Workbook

_____ Study Guide: CRB pages 26–27; Spanish Study Guide

Extending the Lesson

_____ Challenge: SE page 402; CRB page 28

ASSESSMENT OPTIONS

_____ Daily Quiz (8.3): TE page 402 or Transparencies

_____ Standardized Test Practice: SE page 402

Chapter Pacing Guide	
Day	**Lesson**
1	8.1; 8.2 (begin)
2	8.2 (end); **8.3**
3	8.4; 8.5
4	8.6
5	8.7; 8.8
6	8.9
7	Ch. 8 Review and Projects

Notes

_____

_____

_____

Lesson 8.3

Name _____ Date _____

Practice A

For use with pages 398–402

Identify the *x*-intercept and the *y*-intercept of the line.

1.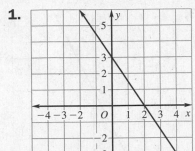

2.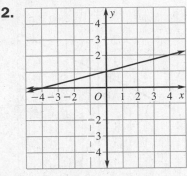

3.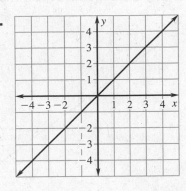

Draw the line with the given intercepts.

4. *x*-intercept: 1
   y-intercept: 4

5. *x*-intercept: 5
   y-intercept: −3

6. *x*-intercept: −3
   y-intercept: −2

7. *x*-intercept: −2
   y-intercept: 4

8. You are kayaking along a 24-mile stretch of river. You travel 6 miles per hour when paddling and 2 miles per hour when drifting. Write and graph an equation describing your possible paddling and drifting times. Give three possible combinations of paddling and drifting times.

9. A fitness center charges $25 for a yoga class and $45 for an aerobic class. The fitness center receives a total of $675 for the two classes. Write and graph an equation describing the possible number of people who joined the yoga and aerobic classes. Give three possible combinations of the number of people in the classes.

Find the intercepts of the equation's graph. Then graph the equation.

10. $3x + y = 6$

11. $-4x + 3y = 12$

12. $-5x + 4y = -20$

13. $2x + y = -4$

14. $5x + 3y = -15$

15. $-x + 7y = -7$

16. $6x + y = -6$

17. $-5x + 8y = 40$

18. $-9x - 4y = -36$

19. At the start of a trip, you fill up your SUV's fuel tank with gas. After you drive for *x* hours, the amount *y* (in gallons) of gas remaining is given by the equation $y = 22 - 3x$.

   a. Find the *x*-intercept and the *y*-intercept of the given equation's graph. Use the intercepts to graph the equation.

   b. What real-life quantities do the *x*- and *y*-intercepts represent in this situation?

   c. After how many hours of driving do you have $\frac{1}{2}$ tank of gas left?

Lesson 8.3

Name _____ Date _____

Practice B

For use with pages 398–402

Identify the x-intercept and the y-intercept of the line.

1.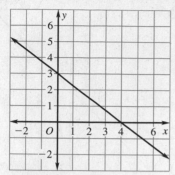

2.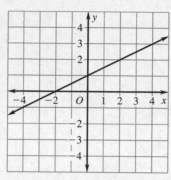

3.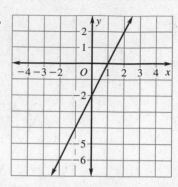

Find the intercepts of the equation's graph. Then graph the equation.

4. $-x + 3y = -9$

5. $2x + 5y = -20$

6. $-3x + 4y = 36$

7. $6x + 7y = 42$

8. $4x + 5y = -60$

9. $2x + y = 14$

10. $-\frac{1}{3}x + \frac{7}{6}y = -\frac{7}{3}$

11. $-\frac{3}{5}x + \frac{1}{5}y = \frac{9}{5}$

12. $\frac{3}{8}x + \frac{1}{2}y = -3$

13. $-21.9x + 6.57y = 65.7$

14. $-8.5x + 13.6y = -68$

15. $-6.5x + 1.3y = 3.25$

16. You are in charge of buying salads for a picnic. You have $20 and plan to buy potato salad and pasta salad. Potato salad costs $1.25 per pound, and pasta salad costs $2.50 per pound. Write an equation describing the possible amounts of potato salad and pasta salad that you can buy. Use intercepts to graph the equation.

17. A car rental agency rents economy and luxury cars by the day. The number of economy cars y rented in one day is given by the equation $y = 24 - 4x$, where x is the number of luxury cars rented. Find the x-intercept and the y-intercept of the given equation's graph. Use the intercepts to graph the equation. How many economy cars are rented when 4 luxury cars have been rented?

18. The rectangle shown has a perimeter of 52 inches.

a. Write an equation describing the possible values of x and y.

b. Use intercepts to graph the equation from part (a).

c. Give three pairs of whole-number values of x and y that could represent side lengths of the rectangle.

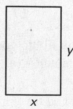

Name _____ Date _____

Practice C

For use with pages 398–402

Identify the *x*-intercept and the *y*-intercept of the line.

1.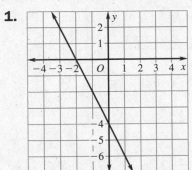

2.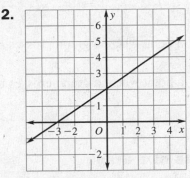

3.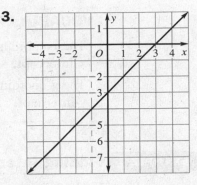

Find the intercepts of the equation's graph. Then graph the equation.

4. $3x + 5y = -45$

5. $-7x + 8y = -56$

6. $y = \frac{6}{11}x + 6$

7. $y = \frac{7}{5}x + 7$

8. $\frac{6}{5}x + \frac{4}{5}y = -\frac{12}{5}$

9. $-\frac{1}{3}x + \frac{5}{9}y = -\frac{10}{3}$

10. $-4.5x + 6y = 7.2$

11. $26.1x + 20.3y = 365.4$

12. $1.4x + 3.5 = -y$

13. $-0.8x + y = -6.8$

14. $y + \frac{18}{5} = \frac{18}{37}x$

15. $y + \frac{94}{29}x = -\frac{47}{5}$

16. A pet grooming store charges $28 for a basic grooming and $42 for a deluxe grooming that includes a shampoo and nail trim. On a certain day, sales at the store total $1162. Write and graph an equation describing the possible number of basic and deluxe groomings that could have been done. Give three possible combinations of basic and deluxe groomings.

17. The isosceles triangle shown has a perimeter of 18 centimeters.

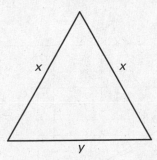

a. Write an equation describing the possible values of *x* and *y*.

b. Use intercepts to graph the equation from part (a).

c. Give three pairs of whole-number values of *x* and *y* that could represent side lengths of the isosceles triangle. Check each ordered pair you list by drawing the triangle with a ruler.

18. A line has an *x*-intercept of 0. What is the *y*-intercept?

19. What kind of line has no *x*-intercept?

Name _____ Date _____

Study Guide

For use with pages 398–402

GOAL Use *x*- and *y*-intercepts to graph linear equations.

> ## VOCABULARY
>
> The *x*-coordinate of a point where a graph crosses the *x*-axis is an **x-intercept.**
>
> The *y*-coordinate of a point where a graph crosses the *y*-axis is a **y-intercept.**

EXAMPLE 1 Finding Intercepts of a Graph

Find the intercepts of the graph of $5x + 9y = 45$.

Solution

To find the *x*-intercept, let $y = 0$ and solve for *x*.

$5x + 9y = 45$	Write original equation.
$5x + 9(0) = 45$	Substitute 0 for *y*.
$5x = 45$	Simplify.
$x = 9$	Divide each side by 5.

To find the *y*-intercept, let $x = 0$ and solve for *y*.

$5x + 9y = 45$	Write original equation.
$5(0) + 9y = 45$	Substitute 0 for *x*.
$9y = 45$	Simplify.
$y = 5$	Divide each side by 9.

Answer: The *x*-intercept is 9, and the *y*-intercept is 5.

EXAMPLE 2 Using Intercepts to Graph a Linear Equation

Graph the equation $5x + 9y = 45$ from Example 1.

The *x*-intercept is 9, so plot the point (9, 0).
The *y*-intercept is 5, so plot the point (0, 5).

Draw a line through the two points.

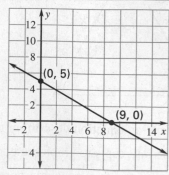

Exercises for Examples 1 and 2

Find the intercepts of the equation's graph. Then graph the equation.

1. $y = 3x + 6$

2. $5x - 3y = 45$

3. $\frac{2}{3}x + y = 4$

Name _____ Date _____

Study Guide

For use with pages 398–402

EXAMPLE 3 **Writing and Graphing an Equation**

For a school fund-raiser, you are selling movie tickets and pizzas. You earn $6 for each packet of movie tickets you sell and $3 for each pizza you sell. You want to raise $300. Write and graph an equation describing your possible sales. Give three possible combinations of ticket packets and pizzas sold.

Solution

(1) To write an equation, let x be the number of movie ticket packets you sell and let y be the number of pizzas you sell. First write a verbal model.

Profit for selling one ticket packet	$\cdot$	Ticket packets sold	$+$	Profit for selling one pizza	$\cdot$	Pizzas sold	$=$	Total profit

Then use the verbal model to write the equation.

$$6x + 3y = 300$$

(2) To graph the equation, find and use the intercepts.

Find x-intercept:
$$6x + 3y = 300$$
$$6x + 3(0) = 300$$
$$6x = 300$$
$$x = 50$$

Find y-intercept:
$$6x + 3y = 300$$
$$6(0) + 3y = 300$$
$$3y = 300$$
$$y = 100$$

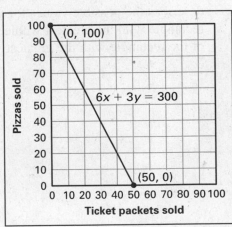

(3) Three points on the graph are (0, 100), (30, 40), and (50, 0). So, you can not sell any ticket packets and sell 100 pizzas, or sell 30 ticket packets and 40 pizzas, or sell 50 ticket packets and not sell any pizzas.

Exercise for Example 3

4. You plan to spend $100 on DVDs and videos. DVDs cost $20 and videos cost $10. Write an equation describing the possible number of DVDs x and videos y you can buy. Then use intercepts to graph the equation. Finally, give three possible combinations of DVDs and videos you can buy.

Lesson 8.3

LESSON
8.3 **Challenge Practice**

For use with pages 398–402

Find the intercepts of the equation's graph. Then graph the equation.

1. $1.2y - 2.4x = 4.8$

2. $\frac{4}{3}x + \frac{1}{2}y = \frac{5}{6}$

3. $\frac{3}{8}x - \frac{1}{5}y = \frac{3}{4}$

4. $3.2 - 0.8y = 0.4x$

5. Draw the graph of an equation whose intercepts are the same number. Identify the intercepts.

6. Draw the graph of an equation whose intercepts are opposite numbers. Identify the intercepts.

7. Write the equation of a line whose graph has an x-intercept of 1 and a positive y-intercept.

8. Write the equation of a line whose graph has a y-intercept of -2 and a positive x-intercept.

9. A rectangle has a perimeter of 12 centimeters. Let x represent the rectangle's width and let y represent the rectangle's length. Graph the equation that describes the possible values for x and y. What are the dimensions of the rectangle with the greatest whole number length? What are the dimensions of the rectangle with the greatest whole number width?

Lesson 8.3

Teacher's Name _____ Class _____ Room _____ Date _____

Lesson Plan

1-day lesson (See *Pacing and Assignment Guide*, TE page 382A)

For use with pages 403–409

GOAL **Find and interpret slopes of lines.**

State/Local Objectives _____

_____

✓ **Check the items you wish to use for this lesson.**

STARTING OPTIONS

____ Homework Check (8.3): TE page 400; Answer Transparencies

____ Homework Quiz (8.3): TE page 402; Transparencies

____ Warm-Up: Transparencies

TEACHING OPTIONS

____ Notetaking Guide

____ Concept Activity: 1–3, SE page 403

____ Examples: 1–4, SE pages 404–406

____ Extra Examples: TE pages 405–406

____ Checkpoint Exercises: 1–8, SE pages 405–406

____ Concept Check: TE page 406

____ Guided Practice Exercises: 1–7, SE page 407

APPLY/HOMEWORK

Homework Assignment

____ Basic: pp. 407–409 Exs. 8–16, 18–27, 35, 41–53

____ Average: pp. 407–409 Exs. 8–24, 28–38, 41–53

____ Advanced: pp. 407–409 Exs. 11–21, 28–42*, 47–53

Reteaching the Lesson

____ Practice: CRB pages 31–33 (Level A, Level B, Level C); Practice Workbook

____ Study Guide: CRB pages 34–35; Spanish Study Guide

Extending the Lesson

____ Challenge: SE page 409; CRB page 36

ASSESSMENT OPTIONS

____ Daily Quiz (8.4): TE page 409 or Transparencies

____ Standardized Test Practice: SE page 409

Notes _____

_____

_____

Teacher's Name _____ Class _____ Room _____ Date _____

Lesson Plan for Block Scheduling

Half-block lesson (See *Pacing and Assignment Guide*, TE page 382A)

For use with pages 403–409

GOAL **Find and interpret slopes of lines.**

State/Local Objectives _____

_____

✓ **Check the items you wish to use for this lesson.**

Chapter Pacing Guide	
Day	**Lesson**
1	8.1; 8.2 (begin)
2	8.2 (end); 8.3
3	**8.4**; 8.5
4	8.6
5	8.7; 8.8
6	8.9
7	Ch. 8 Review and Projects

STARTING OPTIONS

_____ Homework Check (8.3): TE page 400; Answer Transparencies

_____ Homework Quiz (8.3): TE page 402; Transparencies

_____ Warm-Up: Transparencies

TEACHING OPTIONS

_____ Notetaking Guide

_____ Concept Activity: 1–3, SE page 403

_____ Examples: 1–4, SE pages 404–406

_____ Extra Examples: TE pages 405–406

_____ Checkpoint Exercises: 1–8, SE pages 405–406

_____ Concept Check: TE page 406

_____ Guided Practice Exercises: 1–7, SE page 407

APPLY/HOMEWORK

Homework Assignment

_____ Block Schedule: pp. 407–409 Exs. 8–24, 28–38, 43–53 (with 8.5)

Reteaching the Lesson

_____ Practice: CRB pages 31–33 (Level A, Level B, Level C); Practice Workbook

_____ Study Guide: CRB pages 34–35; Spanish Study Guide

Extending the Lesson

_____ Challenge: SE page 409; CRB page 36

ASSESSMENT OPTIONS

_____ Daily Quiz (8.4): TE page 409 or Transparencies

_____ Standardized Test Practice: SE page 409

Notes _____

_____

_____

_____

LESSON
8.4 Practice A

For use with pages 403–409

Tell whether the slope of the line is *positive*, *negative*, *zero*, or *undefined*. Then find the slope.

1.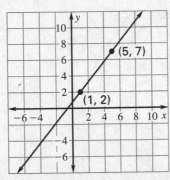

2.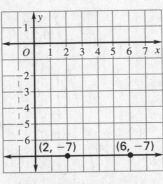

3.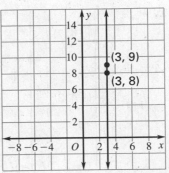

Find the coordinates of two points on the line with the given equation. Then use the points to find the slope of the line.

4. $y = 2x + 6$

5. $y = -3$

6. $y = \frac{5}{4}x - 7$

7. $2x + 3y = 5$

8. $-6x + 5y = 45$

9. $x = 2$

Find the slope of the line through the given points.

10. $(9, 6), (21, 14)$

11. $(8, 0), (10, 10)$

12. $(3, 9), (16, 9)$

13. $(6, 1), (7, -2)$

14. $(-9, -2), (-7, -3)$

15. $(-5, -4), (-5, -7)$

16. $(-4, -8), (-6, -11)$

17. $(-2, -3), (10, 15)$

18. $(7, 6), (12, -14)$

19. $(-13, 6), (8, -17)$

20. $(20, -18), (1, -2)$

21. $(-3, -4), (15, 16)$

22. A ladder is leaning against a house. The bottom of the ladder is 8 feet from the house. The top of the ladder is leaning against the house 15 feet above the ground. Find the slope of the ladder.

23. The graph shows the distance traveled by a car as a function of time.

    a. Find the slope of the line.

    b. What information about the car can you obtain from the slope?

    c. A second car is traveling at 50 miles per hour. Suppose you made a graph showing the distance traveled by the second car as a function of time. How would the graph for the second car compare with the graph of the first car? Explain your thinking.

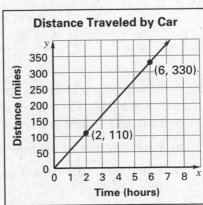

Distance Traveled by Car

Name _____ Date _____

Practice B

For use with pages 403–409

Tell whether the slope of the line is *positive*, *negative*, *zero*, or *undefined*. Then find the slope.

1.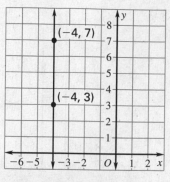

2.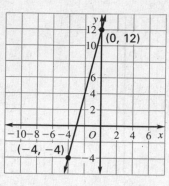

3.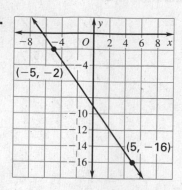

Find the coordinates of two points on the line with the given equation. Then use the points to find the slope of the line.

4. $y = -3x + 11$

5. $y = -17$

6. $y = \frac{7}{8}x - 11$

7. $9x + 8y = 56$

8. $x = 10$

9. $7y - 3x = -147$

Find the slope of the line through the given points.

10. $(6, 3), (14, 19)$

11. $(10, 11), (15, 16)$

12. $(8, 48), (16, 24)$

13. $(1, 5), (36, 19)$

14. $(4, 4), (32, 18)$

15. $(9, 4), (32, 17)$

16. $(-6, -17), (-22, -12)$

17. $(-9, -7), (-11, -13)$

18. $(7, -20), (-13, 10)$

19. $(2, -11), (-13, 14)$

20. $(-4, 15), (-9, 11)$

21. $(4, 4), (14, 10)$

22. The slope of the roof of a house is called the pitch of the roof. Find the pitch of the roof shown.

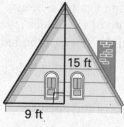

15 ft

9 ft

23. A manufacturing company spent $700 on equipment and then a fixed amount per unit. The graph shows the cost to make *x* units at the manufacturing company.

a. Find the slope of the line.

b. What information about the company can you obtain from the slope?

c. A second manufacturing company spent $700 on equipment and $2.50 per unit. Suppose you made a graph showing the cost to make *x* units at the second manufacturing company. How would the graph of the second company compare with the graph of the first company? Explain your thinking.

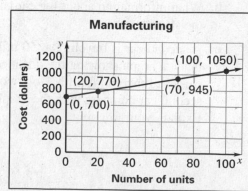

Lesson 8.4

LESSON
8.4 **Practice C**
For use with pages 403–409

Find the coordinates of two points on the line with the given equation. Then use the points to find the slope of the line.

1. $4x + 11y = 25$

2. $y = \frac{1}{3}x - 14$

3. $9x + 25y = 90$

4. $17x + 7y = -89$

5. $5y - 3x = 17$

6. $3y - 8x = 65$

Find the slope of the line through the given points.

7. $(8, 14), (10, 24)$

8. $(10, 44), (24, 16)$

9. $(7, 10), (33, 15)$

10. $(-6, 17), (-7, 5)$

11. $(9, 6), (21, 15)$

12. $(-1, -15), (-26, -10)$

13. $(5, -7), (-19, 9)$

14. $(7, -13), (-33, 19)$

15. $(-1, 11), (-5, 22)$

16. $(-2, -13), (-33, -16)$

17. $(3, 43), (25, 21)$

18. $(10, 1), (33, 24)$

19. The figure shows a person parasailing from a boat on the water. Find the slope of the tow line from the boat to the person parasailing.

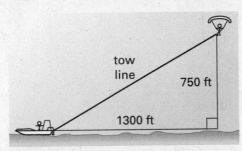

tow line
750 ft
1300 ft

20. The grade of a road is its slope written as a percent. A warning sign must be posted if a section of road has a grade of at least 8% and is more than 750 feet long.

  a. A road rises 65 feet over a horizontal distance of 770 feet. Should a warning sign be posted? Explain your thinking.

  b. The grade of a section of road that stretches over a horizontal distance of 1500 feet is 12%. How many feet does the road rise over that distance?

21. The slope of one line is greater than the slope of a second line. Does this mean that the first line is *steeper* than the second line? Explain your reasoning.

Study Guide

For use with pages 403–409

GOAL Find and interpret slopes of lines.

VOCABULARY

The **slope** of a line is the ratio of the line's vertical change, called the **rise,** to its horizontal change, called the **run.**

EXAMPLE 1 **Finding Slope**

A ladder is leaning on a house. Its base is 6 feet from the house and it is resting on the house at a height of 22 feet. Find its slope.

$$\text{slope} = \frac{\text{rise}}{\text{run}} = \frac{22}{6} = \frac{11}{3}$$

Answer: The ladder has a slope of $\frac{11}{3}$.

22 ft

6 ft

Exercise for Example 1

1. A railing has a rise of 15 feet and a run of 10 feet. Find the slope of the railing.

EXAMPLE 2 **Finding Positive and Negative Slope**

Find the slope of the line shown.

a. $m = \dfrac{\text{rise}}{\text{run}} = \dfrac{\text{difference of } y\text{-coordinates}}{\text{difference of } x\text{-coordinates}}$

$= \dfrac{4 - (-2)}{1 - (-1)}$

$= \dfrac{6}{2}$

$= 3$

Answer: The slope is 3.

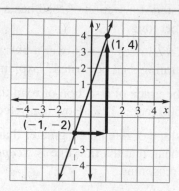

(1, 4)

(−1, −2)

b. $m = \dfrac{\text{rise}}{\text{run}} = \dfrac{\text{difference of } y\text{-coordinates}}{\text{difference of } x\text{-coordinates}}$

$= \dfrac{-4 - (-2)}{3 - 0}$

$= \dfrac{-2}{3}$

$= -\dfrac{2}{3}$

Answer: The slope is $-\dfrac{2}{3}$.

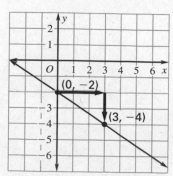

(0, −2)

(3, −4)

Lesson 8.4

Name _____   Date _____

Study Guide

For use with pages 403–409

EXAMPLE 3 **Zero and Undefined Slope**

Find the slope of the line shown.

a.

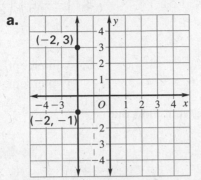

b.

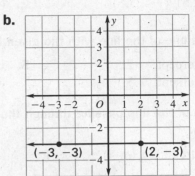

$m = \dfrac{\text{rise}}{\text{run}} = \dfrac{\text{difference of } y\text{-coordinates}}{\text{difference of } x\text{-coordinates}}$

$= \dfrac{3 - (-1)}{-2 - (-2)}$

$= \dfrac{4}{0}$ Division by zero is undefined.

Answer: The slope is undefined.

$m = \dfrac{\text{rise}}{\text{run}} = \dfrac{\text{difference of } y\text{-coordinates}}{\text{difference of } x\text{-coordinates}}$

$= \dfrac{-3 - (-3)}{2 - (-3)}$

$= \dfrac{0}{5} = 0$

Answer: The slope is 0.

EXAMPLE 4 **Interpreting Slope as a Rate of Change**

The graph shows the Calories burned by dancing as a function of time. The slope of the line gives the number of Calories burned per unit of time, which is the *rate of change* in Calories burned with respect to time. Find the Calories burned per minute.

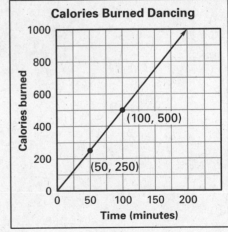

Solution

Use the points (50, 250) and (100, 500) to find the slope of the line.

$m = \dfrac{\text{rise}}{\text{run}} = \dfrac{\text{difference of } y\text{-coordinates}}{\text{difference of } x\text{-coordinates}}$

$= \dfrac{500 \text{ Cal} - 250 \text{ Cal}}{100 \text{ min} - 50 \text{ min}} = \dfrac{250 \text{ Cal}}{50 \text{ min}} = 5 \text{ Cal/min}$

Answer: Dancing burns 5 Calories per minute.

Exercises for Examples 2–4

In Exercises 2–5, find the slope of the line through the given points. Tell whether the slope is *positive*, *negative*, *zero*, or *undefined*.

2. $(-3, 3), (2, -1)$ **3.** $(5, 6), (-3, 6)$ **4.** $(-3, -3), (9, 7)$ **5.** $(3, 1), (3, -5)$

6. You burn 108 Calories playing basketball for 12 minutes and 135 Calories playing basketball for 15 minutes. Draw a graph of Calories burned as a function of time. What information can you obtain from the slope?

Name _____ Date _____

Challenge Practice

For use with pages 403–409

Find the slope of the line that passes through the given points.

1. $(5, 2.5), (-4, 5.5)$

2. $(6.2, -2.1), (7.4, 1.5)$

3. $\left(-\frac{1}{6}, \frac{1}{2}\right), \left(\frac{1}{4}, \frac{1}{3}\right)$

Find the slope of the line with the given intercepts.

4. x-intercept: -2.5
y-intercept: 1.5

5. x-intercept: 3.2
y-intercept: -1.6

6. x-intercept: -0.4
y-intercept: -0.6

The slope of the line passing through the two points is given. Find the value of a.

7. Points: $(3, 2), (a, 4)$; $m = -1$

8. Points: $(a, 1), (-4, 5)$; $m = -2$

9. Points: $(-4, a), (1, 3)$; $m = 2$

10. Without graphing, choose a point P so that the line through $(1, -2)$ and P has the given slope.

a. $\frac{2}{3}$

b. $\frac{3}{2}$

c. $-\frac{2}{3}$

d. $-\frac{3}{2}$

LESSON 8.5

Lesson Plan

1-day lesson (See *Pacing and Assignment Guide*, TE page 382A)

For use with pages 412–417

GOAL Graph linear equations in slope-intercept form.

Teacher's Name _____ Class _____ Room _____ Date _____

State/Local Objectives _____

✓ **Check the items you wish to use for this lesson.**

STARTING OPTIONS

____ Homework Check (8.4): TE page 407; Answer Transparencies

____ Homework Quiz (8.4): TE page 408; Transparencies

____ Warm-Up: Transparencies

TEACHING OPTIONS

____ Notetaking Guide

____ Activity Master: CRB page 39

____ Examples: 1–4, SE pages 412–414

____ Extra Examples: TE pages 413–414

____ Checkpoint Exercises: 1–6, SE pages 413–414

____ Concept Check: TE page 414

____ Guided Practice Exercises: 1–9, SE page 415

____ Technology Keystrokes for Exercise 20(a) on SE page 416: CRB page 40

APPLY/HOMEWORK

Homework Assignment

____ Basic: pp. 415–417 Exs. 10–20, 22–30, 34–36

____ Average: pp. 415–417 Exs. 10–13, 17–24, 28–32, 34–47

____ Advanced: pp. 415–417 Exs. 10–13, 17–21, 25–35*, 38–47

Reteaching the Lesson

____ Practice: CRB pages 41–43 (Level A, Level B, Level C); Practice Workbook

____ Study Guide: CRB pages 44–45; Spanish Study Guide

Extending the Lesson

____ Real-World Problem Solving: CRB page 46

____ Challenge: SE page 417; CRB page 47

ASSESSMENT OPTIONS

____ Daily Quiz (8.5): TE page 417 or Transparencies

____ Standardized Test Practice: SE page 417

____ Quiz (8.1–8.5): SE page 418; Assessment Book page 96

Notes

LESSON
8.5
Teacher's Name _____ Class _____ Room _____ Date _____

Lesson Plan for Block Scheduling
Half-block lesson (See *Pacing and Assignment Guide*, TE page 382A)
For use with pages 412–417

GOAL Graph linear equations in slope-intercept form.

State/Local Objectives _____

✓ Check the items you wish to use for this lesson.

STARTING OPTIONS
____ Homework Check (8.4): TE page 407; Answer Transparencies
____ Homework Quiz (8.4): TE page 408; Transparencies
____ Warm-Up: Transparencies

TEACHING OPTIONS
____ Notetaking Guide
____ Activity Master: CRB page 39
____ Examples: 1–4, SE pages 412–414
____ Extra Examples: TE pages 413–414
____ Checkpoint Exercises: 1–6, SE pages 413–414
____ Concept Check: TE page 414
____ Guided Practice Exercises: 1–9, SE page 415
____ Technology Keystrokes for Exercise 20(a) on SE page 416: CRB page 40

APPLY/HOMEWORK

Homework Assignment
____ Block Schedule: pp. 415–417 Exs. 10–13, 17–24, 28–32, 34–47 (with 8.4)

Reteaching the Lesson
____ Practice: CRB pages 41–43 (Level A, Level B, Level C); Practice Workbook
____ Study Guide: CRB pages 44–45; Spanish Study Guide

Extending the Lesson
____ Real-World Problem Solving: CRB page 46
____ Challenge: SE page 417; CRB page 47

ASSESSMENT OPTIONS
____ Daily Quiz (8.5): TE page 417 or Transparencies
____ Standardized Test Practice: SE page 417
____ Quiz (8.1–8.5): SE page 418; Assessment Book page 96

Notes _____

Chapter Pacing Guide	
Day	**Lesson**
1	8.1; 8.2 (begin)
2	8.2 (end); 8.3
3	8.4; **8.5**
4	8.6
5	8.7; 8.8
6	8.9
7	Ch. 8 Review and Projects

LESSON
8.5 Activity Master
For use before Lesson 8.5

Name _____ Date _____

Goal
Use a graph to investigate properties of equations in slope-intercept form.

Materials
• pencil and paper
• graph paper

Investigating Slope-Intercept Form

In this activity, you will explore the slope-intercept form of linear equations.

INVESTIGATE Graph $y = 3x - 6$. Then identify the slope and the y-intercept.

1 Find the x-intercept.

$y = 3x - 6$
$0 = 3x - 6$
$6 = 3x$
$2 = x$

2 Find the y-intercept.

$y = 3x - 6$
$y = 3(0) - 6$
$y = -6$

3 Use the intercepts to graph the equation. Then identify the slope and the y-intercept.

$m = \dfrac{6}{2} = 3$

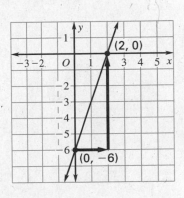

Answer: The line has a slope of 3 and a y-intercept of -6.

DRAW CONCLUSIONS Graph the equation. Then identify the slope and the y-intercept of the graph.

1. $y = -x + 4$ **2.** $y = \dfrac{1}{3}x - 2$ **3.** $y = 6x + 1$

4. Do you see any patterns in Exercises 1−3? Compare the coefficient of x and the constant term of the equation to the slope and the y-intercept.

In Exercises 5–8, use the equation $2x + y - 5 = 0$.

5. Graph the equation and identify the slope and y-intercept.

6. Identify the coefficient of x. Is the slope equivalent to the coefficient of x?

7. Identify the constant term in the equation. Is the y-intercept equivalent to the constant term?

8. Write the equation in function form. Does the pattern established in Exercise 4 hold true for the equation in this form? Explain your results.

9. Find the slope and the y-intercept of $4x - y = 3$ without graphing the equation.

Pre-Algebra
Chapter 8 Resource Book **39**

Name _____ Date _____

LESSON
8.5 Technology Keystrokes

For use with Exercise 21(a), page 416

TI-73 Explorer

21. a. Adjust the window:

[WINDOW] [(−)] 6.75 [ENTER] 16.75 [ENTER] [ENTER] [ENTER]

[(−)] 10 [ENTER] 10

Graph the equation: [Y=] 2000 [−] 250 [x] [GRAPH]

Find the *y*-intercept: [TRACE] (move cursor along the graph using [◀] or [▶])

Name _____ Date _____

Practice A
For use with pages 412–417

1. Without graphing, tell whether the lines $y = \frac{2}{3}x + 1$ and $y = -\frac{3}{2}x + 1$ are *parallel*, *perpendicular*, or *neither*.

Identify the slope and y-intercept of the line with the given equation.

2. $y = -5x$

3. $y = 2x + 1$

4. $y = -4x - 2$

5. $x + y = 5$

6. $2x + 3y = 6$

7. $4x - 2y = 8$

Match the equation with its graph.

8. $y = 2x + 1$

9. $y = x + 2$

10. $y = x - 2$

A.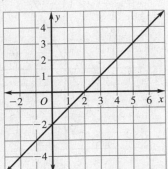

B.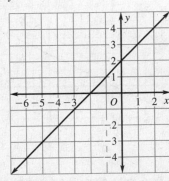

C.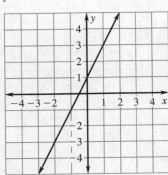

Identify the slope and y-intercept of the line with the given equation. Use the slope and y-intercept to graph the equation.

11. $y = 3x + 2$

12. $y = -\frac{1}{2}x + 3$

13. $y - 2x = 12$

14. $2y - 4x = 12$

For the line with the given equation, find the slope of a parallel line and the slope of a perpendicular line.

15. $y = 6x - 3$

16. $y = \frac{3}{4}x + 4$

17. $y - x = 4$

18. $y + 3x = 7$

19. You fill an aquarium with water. The water is 22 inches deep. The water evaporates at a rate of 2 inches per week.

   a. Write an equation that approximates the depth y of water in the aquarium x weeks after you fill it.

   b. After two weeks, you have not yet refilled the aquarium. What is the depth of the water?

Name _____ Date _____

Practice B
For use with pages 412–417

Identify the slope and *y*-intercept of the line with the given equation.

1. $y = -\frac{1}{3}x + 6$

2. $y = \frac{3}{4}x$

3. $y - 4x = -8$

4. $3x - y = 12$

5. $2x + 6y = 12$

6. $3x + 5y - 15 = 0$

Match the equation with its graph.

7. $y = \frac{1}{2}x + 2$

8. $y = 2x + \frac{1}{2}$

9. $y = -2x + 2$

A.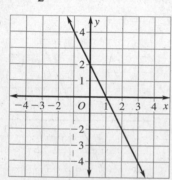

B.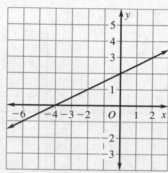

C.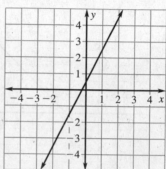

Identify the slope and *y*-intercept of the line with the given equation. Use the slope and *y*-intercept to graph the equation.

10. $y = \frac{5}{4}x + 1$

11. $y - \frac{3}{2}x = 3$

12. $3y + 4x = 24$

13. $x - 3y = 9$

For the line with the given equation, find the slope of a parallel line and the slope of a perpendicular line.

14. $y = 12x - 1$

15. $y = \frac{6}{5}x + 144$

16. $y - 7 = 0$

17. $4y - 4x = 16$

18. $8y + 3x - 32 = 0$

19. $4x + 6y = 9$

20. Forest rangers measure a depth of 82 inches of snow on a mountain peak at 8:00 A.M. Snow is expected to fall at a steady rate of $\frac{3}{4}$ inch per hour throughout the day.

   a. Write an equation that approximates the depth *y* of snow on the mountain peak *x* hours after 8:00 A.M.

   b. The rangers plan to start a controlled avalanche when the depth of snow on the peak reaches 85 inches. At what time will this be?

LESSON 8.5 Practice C

For use with pages 412–417

Identify the slope and *y*-intercept of the line with the given equation. Use the slope and *y*-intercept to graph the equation.

1. $y - 11 = 0$ 　　　**2.** $x + 5 = 1$ 　　　**3.** $-x - y - 13 = 0$

4. $7y - 4x = 4$ 　　　**5.** $12 + 3y = 9x$ 　　　**6.** $15 - 5y + x = 0$

For the line with the given equation, find the slope of a parallel line and the slope of a perpendicular line.

7. $y + 11 = 3$ 　　　**8.** $x - 4 = -3$ 　　　**9.** $11y + 5x = 14$

10. $13y - 8x + 52 = 0$ 　　**11.** $14x + 21y + 8 = 1$ 　　**12.** $12 = 34y - 2x$

In Exercises 13–16, tell whether the lines with the given equations are *parallel*, *perpendicular*, or *neither*.

13. $y - 2x = 8;\ 2y - 4x = 5$ 　　　**14.** $5x - 14y + 12 = 0;\ 10y + 7x = 6$

15. $y = 13;\ x = 5$ 　　　**16.** $y = 7x;\ y = -\dfrac{1}{7}x$

17. You are saving money to buy a guitar by putting $5 into a jar every week. So far, you have $40 in the jar. You need a total of $129 for the guitar.

　　a. Write an equation that represents how much money *y* you will have in the jar after *x* weeks.

　　b. Graph your equation. Tell the meaning of the slope and *y*-intercept.

　　c. How many weeks will you need to save for the guitar?

　　d. Suppose eight weeks ago you had written an equation to represent how much money *y* you would have in the jar after *x* weeks. Describe how the graph of that equation would compare to your graph in part (b).

18. Line *a* passes through the points (0, 8) and (3, 3). Find an equation of line *b* that is perpendicular to line *a* and passes through the point (0, 4).

19. Line *c* passes through the points (3, 7) and (7, 13). Find an equation of line *d* that is parallel to line *c* and passes through the point (0, −4).

8.5 Study Guide

For use with pages 412–417

Name _____ Date _____

GOAL Graph linear equations in slope-intercept form.

VOCABULARY

A linear equation of the form $y = mx + b$ is said to be in **slope-intercept form.** The slope is m and the y-intercept is b.

EXAMPLE 1 **Identifying the Slope and y-Intercept**

Identify the slope and y-intercept of the line with the given equation.

   a. $y = 5 - 4x$ **b.** $2x - 5y = 15$

Solution

  a. Write the equation $y = 5 - 4x$ as $y = -4x + 5$.

    Answer: The line has a slope of -4 and a y-intercept of 5.

  b. Write the equation $2x - 5y = 15$ in slope-intercept form by solving for y.

$2x - 5y = 15$	Write original equation.
$-5y = -2x + 15$	Subtract $2x$ from each side.
$y = \dfrac{2}{5}x - 3$	Multiply each side by $-\dfrac{1}{5}$.
$= \dfrac{2}{5}x + (-3)$	Rewrite $\dfrac{2}{5}x - 3$ as $\dfrac{2}{5}x + (-3)$.

    Answer: The line has a slope of $\dfrac{2}{5}$ and a y-intercept of -3.

EXAMPLE 2 **Graphing an Equation in Slope-Intercept Form**

Graph the equation $y = 2x - 3$.

(1) The y-intercept is -3, so plot the point $(0, -3)$.

(2) The slope is $2 = \dfrac{2}{1}$.

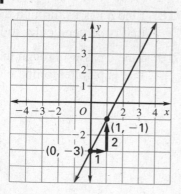

    Starting at $(0, -3)$, plot another point by moving right 1 unit and up 2 units.

(3) Draw a line through the two points.

Exercises for Examples 1 and 2

Identify the slope and y-intercept of the line with the given equation. Use the slope and y-intercept to graph the equation.

  1. $y = -5x + 4$ **2.** $y - x = 1$

  3. $9x + 3y = 10$ **4.** $y + 2 = 4x$

Study Guide
For use with pages 412–417

EXAMPLE 3 Using Slope and y-Intercept in Real Life

You belong to a fitness club. The membership fee is $30 per month. If you are a member, water aerobics classes are $3 per session.

a. Write an equation that gives the total monthly fitness club cost as a function of the number of water aerobics sessions you attend per month.

b. Find the maximum number of sessions you can attend per month and not exceed your $60 monthly fitness budget.

Solution

a. Let x be the number of water aerobics sessions you attend during the month, and let y be the total monthly fitness club cost for that number of sessions. The equation that gives the total monthly fitness club cost as a function of the number of water aerobics sessions attended per month is $y = 3x + 30$.

b. Graph $y = 3x + 30$ on a graphing calculator. Trace along the graph until the cursor is on a point where $y \approx 60$. For this point, $x \approx 10$. So, the maximum number of sessions you can attend is about 10.

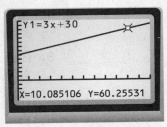

EXAMPLE 4 Finding Slopes of Parallel and Perpendicular Lines

a. Find the slope of a line parallel to $15y - 3x = 5$.

$y = \frac{1}{5}x + \frac{1}{3}$ Write original equation in slope-intercept form.

The slope of the given line is $\frac{1}{5}$. Because parallel lines have the same slope, the slope of a parallel line is also $\frac{1}{5}$.

b. Find the slope of a line perpendicular to $15y - 3x = 5$.

From part (a), the slope of the given line is $\frac{1}{5}$. The slope of a perpendicular line is the negative reciprocal of $\frac{1}{5}$, or -5.

Exercises for Examples 3 and 4

5. Entrance to an amusement park is $20 and games are $3 each. Write an equation that gives the total cost of admission and games. Find the maximum number of games you can play if you have $40.

For the line with the given equation, find the slope of a parallel line and the slope of a perpendicular line.

6. $5y - 10x = 7$ **7.** $11y + 7x = 22$

8. $2y + 5x = 12$ **9.** $y + x = 10$

Name _____ Date _____

LESSON 8.5

Real-World Problem Solving

For use with pages 412–417

Lemonade Stand

Cara is running a lemonade stand during the summer. Each day, she spends about $6 on supplies and sells the lemonade for $.25 per cup. The equation $y = 0.25x - 6$ models the amount of money y Cara makes on a day in which she sells x cups of lemonade.

In Exercises 1–5, use the information above.

1. Find the x- and y-intercepts of the equation. Then determine what the intercepts tell you about the number of cups of lemonade sold and the amount of money Cara makes.

2. What is the slope of the equation? What does it tell you about the amount of money Cara makes?

3. Graph the equation, making sure that you can see both intercepts on the graph.

4. How much money does Cara make selling 20 cups of lemonade? 30 cups of lemonade? 40 cups of lemonade?

5. How many cups of lemonade does Cara have to sell in order to make $1? $10? $100?

Name _____ **Date** _____

Challenge Practice

For use with pages 412–417

Identify the slope and y-intercept of the line with the given equation. Use the slope and y-intercept to graph the equation.

1. $8x - 6y = -5$

2. $0.2y - 1.4x + 0.6 = 0$

3. $\frac{2}{3}x - \frac{1}{6}y - \frac{1}{2} = 0$

4. Graph the equation $3x - 2y = -8$ using its slope and y-intercept. Then, on the same coordinate plane, graph the line whose slope is three times the slope of $3x - 2y = -8$ and that has the same y-intercept.

For the line with the given equation, find the slope of a parallel line and the slope of a perpendicular line.

5. $9x - 4y = -6$

6. $0.6x - 0.3y - 1.2 = 0$

7. $\frac{3}{5}y - \frac{2}{3}x + \frac{7}{15} = 0$

8. For the line that has an x-intercept of -4 and a y-intercept of -3, find the slope of a parallel line and the slope of a perpendicular line.

9. For the line that passes through $(-4, 7)$ and $(3, -5)$, find the slope of a parallel line and the slope of a perpendicular line.

Teacher's Name _____ Class _____ Room _____ Date _____

Lesson Plan

2-day lesson (See *Pacing and Assignment Guide*, TE page 382A)

For use with pages 419–425

GOAL **Write linear equations.**

State/Local Objectives _____

_____

✓ **Check the items you wish to use for this lesson.**

STARTING OPTIONS

_____ Homework Check (8.5): TE page 415; Answer Transparencies

_____ Homework Quiz (8.5): TE page 417; Transparencies

_____ Warm-Up: Transparencies

TEACHING OPTIONS

_____ Notetaking Guide

_____ Examples: Day 1: 1, 3, SE pages 419–420; Day 2: 4–5, SE pages 419–421

_____ Extra Examples: TE pages 420–421

_____ Checkpoint Exercises: Day 1: 1, SE page 419; Day 2: 2–3, SE pages 419–420

_____ Concept Check: TE page 421

_____ Guided Practice Exercises: Day 1: 6, SE page 442; Day 2: 1–5, 7, SE page 422

_____ Technology Activity: SE page 425

APPLY/HOMEWORK

Homework Assignment

_____ Basic: Day 1: pp. 422–424 Exs. 8–11, 18–23, 27–30, 35–42
Day 2: pp. 422–424 Exs. 12–17, 24–26, 43–48

_____ Average: Day 1: pp. 422–424 Exs. 8–11, 18–23, 27–30, 35–42
Day 2: pp. 422–424 Exs. 12–17, 24–26, 31–33, 43–48

_____ Advanced: Day 1: pp. 422–424 Exs. 10, 11, 18–23, 29–31, 40–44;
Day 2: pp. 422–424 Exs. 13–17, 24–26, 32–36*, 45–48

Reteaching the Lesson

_____ Practice: CRB pages 50–52 (Level A, Level B, Level C); Practice Workbook

_____ Study Guide: CRB pages 53–54; Spanish Study Guide

Extending the Lesson

_____ Challenge: SE page 424; CRB page 55

ASSESSMENT OPTIONS

_____ Daily Quiz (8.6): TE page 424 or Transparencies

_____ Standardized Test Practice: SE page 424

Notes _____

Teacher's Name _____ Class _____ Room _____ Date _____

Lesson Plan for Block Scheduling

1-block lesson (See *Pacing and Assignment Guide*, TE page 382A)

For use with pages 419–425

GOAL **Write linear equations.**

State/Local Objectives _____

_____

✓ **Check the items you wish to use for this lesson.**

Chapter Pacing Guide	
Day	Lesson
1	8.1; 8.2 (begin)
2	8.2 (end); 8.3
3	8.4; 8.5
4	**8.6**
5	8.7; 8.8
6	8.9
7	Ch. 7 Review and Projects

STARTING OPTIONS

____ Homework Check (8.5): TE page 415; Answer Transparencies

____ Homework Quiz (8.5): TE page 417; Transparencies

____ Warm-Up: Transparencies

TEACHING OPTIONS

____ Notetaking Guide

____ Examples: 1–5, SE pages 419–421

____ Extra Examples: TE pages 420–421

____ Checkpoint Exercises: 1–3, SE pages 419–420

____ Concept Check: TE page 421

____ Guided Practice Exercises: 1–7, SE page 422

____ Technology Activity: SE page 425

APPLY/HOMEWORK

Homework Assignment

____ Block Schedule: pp. 422–424 Exs. 8–33, 35–48

Reteaching the Lesson

____ Practice: CRB pages 50–52 (Level A, Level B, Level C); Practice Workbook

____ Study Guide: CRB pages 53–54; Spanish Study Guide

Extending the Lesson

____ Challenge: SE page 424; CRB page 55

ASSESSMENT OPTIONS

____ Daily Quiz (8.6): TE page 424 or Transparencies

____ Standardized Test Practice: SE page 424

Notes

_____

_____

_____

Name _____

Date _____

Practice A

For use with pages 419–425

1. Complete the statement with *always*, *sometimes*, or *never*: To draw a line that appears to best fit a set of data points, you should __?__ draw the line through two of the data points.

Write an equation of the line with the given slope and y-intercept.

2. slope = 1; *y*-intercept = 5

3. slope = −2; *y*-intercept = 8

4. slope = 3; *y*-intercept = −10

5. slope = −1; *y*-intercept = −7

Write an equation of the line.

6.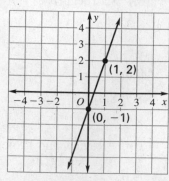

7.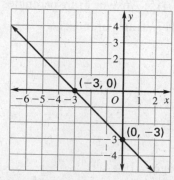

8.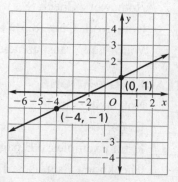

In Exercises 9–11, write an equation of the line through the given points.

9. $(0, 6), (1, 4)$

10. $(0, 9), (−5, 5)$

11. $(−3, −2), (0, 10)$

12. Write an equation of the line that is parallel to the line $y = −3x + 2$ and passes through the point $(0, 3)$.

13. Write an equation of the line that is perpendicular to the line $y = −\frac{3}{4}x − 1$ and passes through the point $(0, −7)$.

14. Show that the table represents a function. Then write an equation for the function.

x	−1	0	1	2	3
y	4	2	0	−2	−4

15. The table shows the number of scientists that attended an annual seminar from 1999–2003.

Years since 1999, x	0	1	2	3	4
Number in attendance, y	463	493	536	574	603

a. Make a scatter plot of the data pairs. Draw the line that appears to best fit the points.

b. Write an equation of your line.

c. Use your equation to predict the number of scientists that will attend in 2008.

LESSON
8.6 **Practice B**
For use with pages 419–425

Write an equation of the line with the given slope and y-intercept.

1. slope $= -3$; y-intercept $= -2$

2. slope $= 5$; y-intercept $= 7$

3. slope $= -\dfrac{3}{4}$; y-intercept $= 3$

4. slope $= \dfrac{5}{2}$; y-intercept $= -6$

Write an equation of the line.

5.

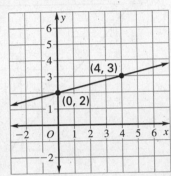

6.

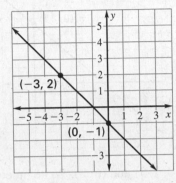

7.

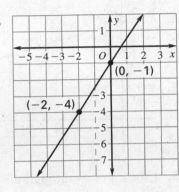

In Exercises 8–10, write an equation of the line through the given points.

8. $(0, 4), (3, 3)$

9. $(2, -3), (0, 5)$

10. $(0, -2), (3, -2)$

In Exercises 11 and 12, use the graph at the right.

11. Write an equation of the line that is parallel to line a and passes through the point $(0, 9)$.

12. Write an equation of the line perpendicular to line a that passes through the point $(0, -2)$.

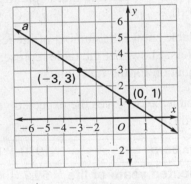

13. Show that the table represents a function. Then write an equation for the function.

x	−4	−2	0	2	4
y	5	4	3	2	1

14. The table shows a boy's height measured each birthday from age 9 until age 13.

Years since age 9, x	0	1	2	3	4
Height (cm), y	133	139	143	149	156

    a. Make a scatter plot of the data pairs. Draw the line that appears to best fit the points.

    b. Write an equation of your line.

    c. Use your equation to predict the boy's height at age 17.

    d. Use your equation to approximate the boy's height on his 8th birthday to the nearest centimeter.

Name _____

Date _____

Write an equation of the line with the given slope and y-intercept.

1. slope $= -\dfrac{5}{3}$; y-intercept $= -5$

2. slope $= \dfrac{8}{5}$; y-intercept $= -12$

Write an equation of the line.

3.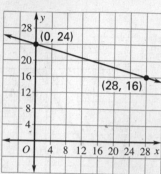

4.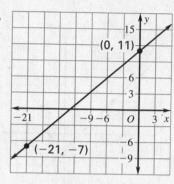

5.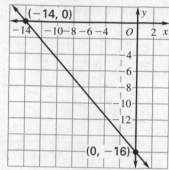

6. Line a passes through the points $(0, -5)$ and $(6, 2)$. Write an equation of the line that is perpendicular to line a and passes through the point $(0, 4)$.

7. Line b passes through the origin and through the point $\left(4, -\dfrac{3}{2}\right)$. Write an equation of the line that is parallel to line b and passes through the point $(0, -7)$.

8. The table shows the number of inches y that are equivalent to x meters. Show that the table represents a function. Then write an equation for the function.

x	0	0.5	4	7
y	0	19.685	157.48	275.59

9. The table below gives the expected years of life for a person given their year of birth.

Year born	1930	1940	1950	1960	1970	1980	1990	2000
Expected years of life	59.7	62.9	68.2	69.7	70.8	73.7	75.4	76.9

a. Let x be the number of years since 1900. Let y be the expected years of life. Make a scatter plot of the data pairs (x, y). Draw the line that appears to best fit the data points.

b. Write an equation of your line.

c. Estimate the year when life expectancy will be 80.

10. The table shows a girl's weight measured each birthday from age 2 until age 7. Use the line of best fit to predict the girl's weight on her 9th birthday to the nearest pound.

Years since age 2, x	0	1	2	3	4	5
Weight (pounds), y	23	25	28	32	35	39

11. In Exercise 10, is the line of best fit appropriate to accurately predict the girl's weight when she turns 25 years old? Explain your reasoning.

Name _____ Date _____

LESSON 8.6 Study Guide

For use with pages 419–425

GOAL Write linear equations.

VOCABULARY

Often, there is no single line that passes through all the points in a data set. In such cases, you can find the **best-fitting line,** which is the line that lies as close as possible to the data points.

EXAMPLE 1 **Writing an Equation Given the Slope and y-Intercept**

Write an equation of the line with a slope of -13 and a y-intercept of 19.

$y = mx + b$ Write general slope-intercept equation.

$y = -13x + 19$ Substitute -13 for m and 19 for b.

EXAMPLE 2 **Writing an Equation of a Graph**

Write an equation of the line shown.

(1) Find the slope m using the labeled points.

$$m = \frac{-2 - (-3)}{2 - 0} = \frac{1}{2}$$

(2) Find the y-intercept b. The line crosses the y-axis at $(0, -3)$, so $b = -3$.

(3) Write an equation of the form $y = mx + b$.

$$y = \frac{1}{2}x - 3$$

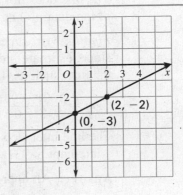

EXAMPLE 3 **Writing Equations of Parallel or Perpendicular Lines**

a. Write an equation of the line that is parallel to the line $y = -15x + 11$ and passes through the point $(0, -8)$.

b. Write an equation of the line that is perpendicular to the line $y = -9x + 6$ and passes through the point $(0, 1)$.

Solution

a. The slope of the given line is -15, so the slope of the parallel line is also -15. The parallel line passes through $(0, -8)$, so its y-intercept is -8.

Answer: An equation of the line is $y = -15x + (-8)$, or $y = -15x - 8$.

b. Because the slope of the given line is -9, the slope of the perpendicular line is the negative reciprocal of -9, or $\frac{1}{9}$. The perpendicular line passes through $(0, 1)$, so its y-intercept is 1.

Answer: An equation of the line is $y = \frac{1}{9}x + 1$.

Name _____ Date _____

Study Guide
For use with pages 419–425

EXAMPLE 4 **Approximating a Best-Fitting Line**

The table shows the average daily cost to community hospitals per patient for the years 1994–2000.

Years since 1994, x	0	1	2	3	4	5	6
Average daily cost	\$931	\$968	\$1006	\$1033	\$1067	\$1103	\$1149

a. Approximate the equation of the best-fitting line for the data.

b. Predict the average daily cost per patient in 2004.

Solution

a. *First*, make a scatter plot of the data pairs.

Next, draw the line that appears to best fit the data points. There should be about the same number of points above the line as below it. The line does not have to pass through any of the data points.

Finally, write an equation of the line. To find the slope, estimate the coordinates of two points on the line, such as (0, 931) and (2, 1001).

$$m = \frac{1001 - 931}{2 - 0} = \frac{70}{2} = 35$$

The line intersects the y-axis at (0, 931), so the y-intercept is 931.

Answer: An approximate equation of the best-fitting line is $y = 35x + 931$.

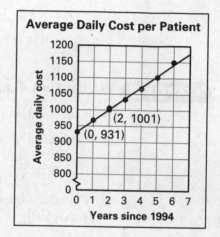

Average Daily Cost per Patient

b. Note that $2004 - 1994 = 10$, so 2004 is 10 years after 1994.

Calculate y when $x = 10$ using the equation from part (a).

$y = 35x + 931$	Write equation of best-fitting line.
$y = 35(10) + 931$	Substitute 10 for x.
$y = 1281$	Simplify.

Answer: In 2004, the average daily cost per patient will be about \$1281.

Exercises for Examples 1–4

Write an equation of the line with the given characteristics.

1. slope $= 7$; y-intercept $= -8$

2. passes through the points $(0, -4)$ and $(-7, 1)$

3. perpendicular to the line $y = \frac{8}{3}x - 5$; passes through the point $(0, 4)$

4. Write the equation of the line that appears to best fit the data points. Use it to predict the value of y when $x = -9$.

x	−3	−2	−1	2	4	5
y	−2	−3	−2	0	1	2

LESSON
8.6

Challenge Practice

For use with pages 419–425

Write an equation of the line that passes through the given points.

1. $\left(\frac{1}{3}, \frac{1}{2}\right), \left(0, \frac{5}{2}\right)$

2. $\left(0, -\frac{8}{3}\right), \left(-\frac{7}{2}, \frac{1}{4}\right)$

3. $(0.8, 4.3), (0, -2.1)$

Write an equation of the line that is parallel to the given line and passes through the given point.

4. Given line: Passes through $(3, 5)$ and $(0, -1)$; Given point: $\left(0, \frac{3}{2}\right)$

5. Given line: Passes through $(0, -5)$ and $(3, -7)$; Given point: $(0, 4)$

Write an equation of the line that is perpendicular to the given line and passes through the given point.

6. Given line: Passes through $(0, -6)$ and $(8, -1)$; Given point: $(0, -2)$

7. Given line: Passes through $(0, 2)$ and $(6, -6)$; Given point: $(0, -3)$

8. Determine whether the line passing through $(3, 6)$ and $(0, 2)$ is parallel, perpendicular, or neither parallel nor perpendicular to the line passing through $(0, -1)$ and $(8, 5)$.

9. Write an equation of the line passing through $(-3, 4)$ and $(2, -1)$. Describe the method you used to determine the equation.

Teacher's Name _____ Class _____ Room _____ Date _____

Lesson Plan

1-day lesson (See *Pacing and Assignment Guide*, TE page 382A)
For use with pages 426–430

GOAL **Use function notation.**

State/Local Objectives _____

_____

✓ **Check the items you wish to use for this lesson.**

STARTING OPTIONS

_____ Homework Check (8.6): TE page 422; Answer Transparencies

_____ Homework Quiz (8.6): TE page 424; Transparencies

_____ Warm-Up: Transparencies

TEACHING OPTIONS

_____ Notetaking Guide

_____ Examples: 1–4, SE pages 426–428

_____ Extra Examples: TE pages 427–428

_____ Checkpoint Exercises: 1–10, SE pages 426–427

_____ Concept Check: TE page 428

_____ Guided Practice Exercises: 1–10, SE page 428

_____ Technology Keystrokes for Exercise 30(a) on SE page 430: CRB page 58

APPLY/HOMEWORK

Homework Assignment

_____ Basic: pp. 429–430 Exs. 11–22, 26–28, 32–43

_____ Average: pp. 429–430 Exs. 14–15, 28–30, 32–44

_____ Advanced: pp. 429–430 Exs. 14–16, 20–33*, 34–37, 40–44

Reteaching the Lesson

_____ Practice: CRB pages 59–61 (Level A, Level B, Level C); Practice Workbook

_____ Study Guide: CRB pages 62–63; Spanish Study Guide

Extending the Lesson

_____ Challenge: SE page 430; CRB page 64

ASSESSMENT OPTIONS

_____ Daily Quiz (8.7): TE page 430 or Transparencies

_____ Standardized Test Practice: SE page 430

Notes _____

_____

_____

LESSON
8.7 Lesson Plan for Block Scheduling
Half-block lesson (See *Pacing and Assignment Guide*, TE page 382A)
For use with pages 426–430

GOAL **Use function notation.**

State/Local Objectives _____

_____

✓ **Check the items you wish to use for this lesson.**

Chapter Pacing Guide	
Day	**Lesson**
1	8.1; 8.2 (begin)
2	8.2 (end); 8.3
3	8.4; 8.5
4	8.6
5	**8.7**; 8.8
6	8.9
7	Ch. 8 Review and Projects

STARTING OPTIONS
____ Homework Check (8.6): TE page 422; Answer Transparencies
____ Homework Quiz (8.6): TE page 424; Transparencies
____ Warm-Up: Transparencies

TEACHING OPTIONS
____ Notetaking Guide
____ Examples: 1–4, SE pages 426–428
____ Extra Examples: TE pages 427–428
____ Checkpoint Exercises: 1–10, SE pages 426–427
____ Concept Check: TE page 428
____ Guided Practice Exercises: 1–10, SE page 428
____ Technology Keystrokes for Exercise 30(a) on SE page 430: CRB page 58

APPLY/HOMEWORK
Homework Assignment
____ Block Schedule: pp. 429–430 Exs. 14–25, 28–30, 32–44 (with 8.8)

Reteaching the Lesson
____ Practice: CRB pages 59–61 (Level A, Level B, Level C); Practice Workbook
____ Study Guide: CRB pages 62–63; Spanish Study Guide

Extending the Lesson
____ Challenge: SE page 430; CRB page 64

ASSESSMENT OPTIONS
____ Daily Quiz (8.7): TE page 430 or Transparencies
____ Standardized Test Practice: SE page 430

Notes _____

_____

_____

_____

Lesson 8.7

Name _____ Date _____

Technology Keystrokes
For use with Exercise 30(a), page 430

TI-73 Explorer

30. a. [Y=] [(−)] 0.117 [x] [+] 1.68 [GRAPH]

LESSON
8.7 Practice A
For use with pages 426–430

1. Write the equation $x + y = 6$ in function form. Then write the function form of the equation in function notation.

2. Suppose f is a linear function where $f(3) = 4$ and $f(8) = 4$. What kind of line is the graph of f?

Let $f(x) = -4x - 6$. Find the indicated value.

3. $f(x)$ when $x = 3$

4. $f(x)$ when $x = -2$

5. x when $f(x) = -6$

6. x when $f(x) = -10$

7. $f(-3)$

8. $f(5)$

Match the function with its graph.

9. $f(x) = x + 2$

10. $f(x) = x - 2$

11. $f(x) = 2x + 1$

A.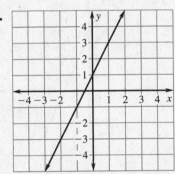

B.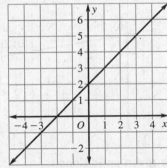

C.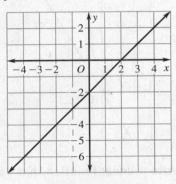

Graph the function.

12. $f(x) = -x + 1$

13. $f(x) = 3x - 2$

14. $f(x) = \frac{3}{4}x - 2$

Write a linear function that satisfies the given conditions.

15. $f(0) = -3, f(3) = 3$

16. $f(0) = 5, f(10) = 0$

17. $g(-4) = 1, g(0) = -2$

18. $h(-8) = 4, h(0) = 28$

19. Maury and Sandra are sharing a ride home from college. The trip is 280 miles. They plan to drive at an average rate of 60 miles per hour.

    a. Use function notation to write an equation giving the distance traveled d (in miles) as a function of the time t (in hours).

    b. How long does it take Maury and Sandra to make the drive?

Lesson 8.7

Name _____ Date _____

Let $f(x) = 4x - 3$ and $h(x) = -5x + 7$. **Find the indicated value.**

1. $f(x)$ when $x = 6$ **2.** $h(x)$ when $x = -5$ **3.** x when $f(x) = -15$

4. x when $h(x) = -13$ **5.** $f(-3) + h(2)$ **6.** $f(5) - h(0)$

Graph the function.

7. $g(x) = 9x - 7$ **8.** $h(x) = -\dfrac{4}{5}x + 1$ **9.** $f(x) = \dfrac{2}{7}x - 3$

Write a linear function that represents the graph.

10.
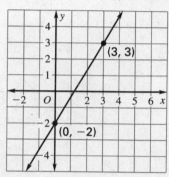

11.
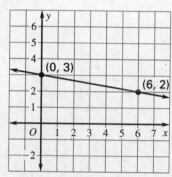

12.
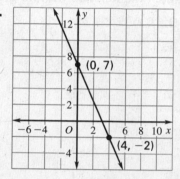

Write a linear function that satisfies the given conditions.

13. $f(0) = 40, f(30) = 65$ **14.** $f(-7) = 8, f(0) = 12$

15. $d(-13) = -9, d(0) = -2$ **16.** $g(0) = 111, g(25) = 286$

17. A PVC (polyvinylchloride) recycling plant uses recent technology to separate PVC from scrap by dissolving the PVC. By 2004, the plant had recycled a total of 20,000 metric tons of PVC. The plant recycles about 8500 metric tons per year. Let t be the number of years since 2004. Use function notation to write an equation giving the total amount of PVC recycled by the plant as a function of t.

18. Currently, there are 4120 gallons of water in Alexa's swimming pool. When filled to the recommended level, the pool holds 4550 gallons. Using a garden hose, she adds 6 gallons of water per minute to the pool.

     a. Use function notation to write an equation giving the amount of water in the pool as a function of the number of minutes x that Alexa runs the hose.

     b. How long will it take Alexa to fill the pool?

LESSON 8.7 Practice C

For use with pages 426–430

Let $d(x) = 11x + 22$ and $g(x) = -9x - 14$. Find the indicated value.

1. $d(x)$ when $x = 8$

2. $g(x)$ when $x = 7$

3. x when $d(x) = 66$

4. x when $g(x) = -41$

5. $d(-5) + g(-5)$

6. $d(6) - 2[g(1)]$

Graph the function.

7. $f(x) = -\dfrac{3}{4}x - 12$

8. $g(x) = \dfrac{10}{3}x - 9$

9. $h(x) = 10x + 120$

Write a linear function that satisfies the given conditions.

10. $f(0) = 36, f(27) = 15$

11. $f(0) = \dfrac{7}{2}, f(9) = 16$

12. $f(1) = 17$, slope = 12

13. $f(3) = 8$, slope = $\dfrac{5}{2}$

14. Use function notation to write two different linear equations whose graphs each have a y-intercept of 15.

15. Write three linear functions $a(x)$, $b(x)$, and $c(x)$ that represent the lines containing the sides of the triangle shown.

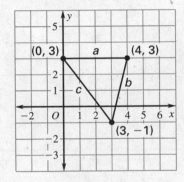

16. The functions $f(x) = 2x - 4$ and $g(x) = -x + 5$ represent lines in the coordinate plane.

   a. Graph both of the functions in one coordinate plane.

   b. Record the coordinates of the point where the graphs intersect.

   c. Check your answer by substituting the coordinates of the point into each function. If the coordinates are correct, each equation should produce a true statement.

17. The points $A(0, 3)$ and $B(4, 6)$ are two vertices of a family of triangles that have a right angle at point A and whose third vertex is point C. Write a function representing the line that contains points A and C. Then use the function to find a pair of coordinates for point C.

Name _____

Date _____

Study Guide

For use with pages 426–430

GOAL Use function notation.

> **VOCABULARY**
>
> When you use an equation to represent a function, it is often convenient to give the function a name, such as f or g. For instance, the function $y = x + 2$ can be written in **function notation** as follows: $f(x) = x + 2$. The symbol $f(x)$, which replaces y, is read "f of x" and represents the value of the function f at x.

EXAMPLE 1 **Working with Function Notation**

Let $f(x) = 5x - 12$. Find $f(x)$ when $x = 2$, and find x when $f(x) = 18$.

a. $f(x) = 5x - 12$ Write function.

    $f(2) = 5(2) - 12$ Substitute 2 for x.

        $= -2$ Simplify.

Answer: When $x = 2$, $f(x) = -2$.

b. $f(x) = 5x - 12$ Write function.

     $18 = 5x - 12$ Substitute 18 for $f(x)$.

     $30 = 5x$ Add 12 to each side.

      $6 = x$ Divide each side by 5.

Answer: When $f(x) = 18$, $x = 6$.

EXAMPLE 2 **Graphing a Function**

Graph the function $f(x) = 9x - 6$.

(1) Rewrite the function as $y = 9x - 6$.

(2) The y-intercept is -6, so plot the point $(0, -6)$.

(3) The slope is 9. Starting at $(0, -6)$, plot another point by moving right 1 unit and up 9 units.

(4) Draw a line through the two points.

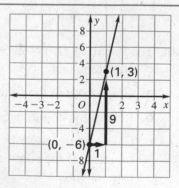

Exercises for Examples 1 and 2

In Exercises 1 and 2, let $f(x) = -2x + 5$. Find the indicated value.

1. $f(x)$ when $x = 1$

2. x when $f(x) = 39$

3. Graph $f(x)$.

Study Guide

For use with pages 426–430

Lesson 8.7

EXAMPLE 3 Writing a Function

Write a linear function h given that $h(0) = -14$ and $h(-3) = -5$.

(1) Find the slope m of the function's graph. From the values of $h(0)$ and $h(-3)$, you know that the graph of h passes through the points $(0, -14)$ and $(-3, -5)$. Use these points to calculate the slope.

$$m = \frac{-5 - (-14)}{-3 - 0} = \frac{9}{-3} = -3$$

(2) Find the y-intercept b of the function's graph. The graph passes through $(0, -14)$, so $b = -14$.

(3) Write an equation of the form $h(x) = mx + b$.

$$h(x) = -3x + (-14) = -3x - 14$$

EXAMPLE 4 Using Function Notation in Real Life

Your car needs new parts. The parts cost $95 and the mechanic charges $50 per hour for labor.

  a. Use function notation to write an equation giving the total repair cost as a function of the hours of labor.

  b. How long did it take the mechanic to install the new parts if the total repair cost is $220?

Solution

  a. Let h be the hours of labor and let $c(h)$ be the total repair cost for h hours of labor. Write a verbal model. Then use the verbal model to write an equation.

Total cost	=	Cost per hour for labor	·	Hours of labor	+	Cost of parts
$c(h)$	=		50h		+	95

  b. Find the value of h for which $c(h) = 220$.

  $c(h) = 50h + 95$ Write function for total cost.

  $220 = 50h + 95$ Substitute 220 for $c(h)$.

  $2.5 = h$ Solve for h.

  Answer: It took the mechanic 2.5 hours to install the new parts.

Exercises for Examples 3 and 4

In Exercises 4 and 5, write a linear function that satisfies the given conditions.

4. $f(0) = 10, f(3) = 7$ **5.** $g(0) = -8, g(9) = -14$

6. You buy gasoline and one gallon of windshield wiper fluid at a gas station. Gasoline is $1.50 per gallon and wiper fluid is $1 per gallon. Use function notation to write an equation giving the total cost c for your stop at the gas station as a function of the number of gallons of gas g you buy. Then find the number of gallons of gas you buy if your total cost for gas and wiper fluid is $13.

Name _____ Date _____

Challenge Practice

For use with pages 426–430

Let $f(x) = -\frac{3}{2}x - 5$. **Find the indicated value.**

1. $f(x)$ when $x = -14$

2. x when $f(x) = -11$

3. $f(x)$ when $x = \frac{4}{9}$

4. x when $f(x) = \frac{7}{8}$

Write a linear function that satisfies the given conditions.

5. $f(0) = -4.8$
$f(1.2) = -0.6$

6. $g\left(\frac{5}{6}\right) = -\frac{3}{8}$
$g(0) = \frac{2}{3}$

7. $h(0) = \frac{3}{4}$
$h(-5) = -\frac{7}{3}$

8. Write a linear function g whose graph passes through $(-2, 2)$ and is perpendicular to the graph of $f(x) = -\frac{3}{2}x + 5$.

9. The length of a rectangle is two more than three times its width. Write a function $P(w)$ for the perimeter of the rectangle in terms of its width. Then use your function to find the perimeter of a rectangle that is 4.5 units wide.

Teacher's Name _____ Class _____ Room _____ Date _____

Lesson Plan

1-day lesson (See *Pacing and Assignment Guide*, TE page 382A)

For use with pages 431–435

GOAL **Graph and solve systems of linear equations.**

State/Local Objectives _____

_____

✓ **Check the items you wish to use for this lesson.**

STARTING OPTIONS

_____ Homework Check (8.7): TE page 429; Answer Transparencies

_____ Homework Quiz (8.7): TE page 430; Transparencies

_____ Warm-Up: Transparencies

TEACHING OPTIONS

_____ Notetaking Guide

_____ Examples: 1–4, SE pages 431–433

_____ Extra Examples: TE pages 432–433

_____ Checkpoint Exercises: 1–4, SE page 432

_____ Concept Check: TE page 433

_____ Guided Practice Exercises: 1–6, SE page 433

_____ Technology Keystrokes for Exs. 25(b) and 29(b) on SE page 435: CRB page 67

APPLY/HOMEWORK

Homework Assignment

_____ Basic: pp. 434–435 Exs. 7–18, 22–24, 30–41

_____ Average: pp. 434–435 Exs. 7–12, 16–28, 32–41

_____ Advanced: pp. 434–435 Exs. 7–12, 16–18, 22–31*, 34–41

Reteaching the Lesson

_____ Practice: CRB pages 68–70 (Level A, Level B, Level C); Practice Workbook

_____ Study Guide: CRB pages 71–72; Spanish Study Guide

Extending the Lesson

_____ Challenge: SE page 435; CRB page 73

ASSESSMENT OPTIONS

_____ Daily Quiz (8.8): TE page 435 or Transparencies

_____ Standardized Test Practice: SE page 435

Notes _____

_____

Teacher's Name _____ Class _____ Room _____ Date _____

Lesson Plan for Block Scheduling

Half-block lesson (See *Pacing and Assignment Guide*, TE page 382A)

For use with pages 431–435

GOAL **Graph and solve systems of linear equations.**

State/Local Objectives _____

_____

✓ **Check the items you wish to use for this lesson.**

Chapter Pacing Guide	
Day	**Lesson**
1	8.1; 8.2
2	8.2 (end); 8.3
3	8.4; 8.5
4	8.6
5	8.7; **8.8**
6	8.9
7	Ch. 8 Review and Projects

STARTING OPTIONS

_____ Homework Check (8.7): TE page 429; Answer Transparencies

_____ Homework Quiz (8.7): TE page 430; Transparencies

_____ Warm-Up: Transparencies

TEACHING OPTIONS

_____ Notetaking Guide

_____ Examples: 1–4, SE pages 431–433

_____ Extra Examples: TE pages 432–433

_____ Checkpoint Exercises: 1–4, SE page 432

_____ Concept Check: TE page 433

_____ Guided Practice Exercises: 1–6, SE page 433

_____ Technology Keystrokes for Exs. 25(b) and 29(b) on SE page 435: CRB page 67

APPLY/HOMEWORK

Homework Assignment

_____ Block Schedule: pp. 434–435 Exs. 7–12, 16–28, 32–41 (with 8.7)

Reteaching the Lesson

_____ Practice: CRB pages 68–70 (Level A, Level B, Level C); Practice Workbook

_____ Study Guide: CRB pages 71–72; Spanish Study Guide

Extending the Lesson

_____ Challenge: SE page 435; CRB page 73

ASSESSMENT OPTIONS

_____ Daily Quiz (8.8): TE page 435 or Transparencies

_____ Standardized Test Practice: SE page 435

Notes _____

_____

_____

LESSON 8.8 Technology Keystrokes

For use with Exercise 25(b) and 29(b), page 435

TI-73 Explorer

25. b. Adjust the window:

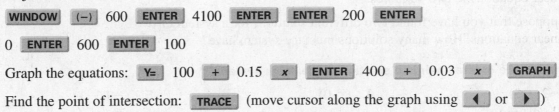

WINDOW [(−)] 600 ENTER 4100 ENTER ENTER 200 ENTER

0 ENTER 600 ENTER 100

Graph the equations: [Y=] 100 [+] 0.15 [x] ENTER 400 [+] 0.03 [x] GRAPH

Find the point of intersection: [TRACE] (move cursor along the graph using [◄] or [►])

29. b. Enter the equations into the graphing calculator: [Y=] 2 [x] ENTER 450 [÷] [x]

Adjust the table setup: [2nd] [TBLSET] 0 ENTER 1

Make a table of solutions for each equation: [2nd] [TABLE]

(press [▼] repeatedly to find the coordinates that satisfy both equations)

Lesson 8.8

LESSON
8.8 Practice A

For use with pages 431–435

1. Describe the numbers of solutions that are possible for a system of two linear equations in two variables.

2. Suppose that you have found two different solutions for a system of two linear equations. How many solutions must the system have?

Tell whether the ordered pair is a solution of the linear system.

3. $(-2, -3)$
$y = 3x + 3$
$y = -x - 5$

4. $(2, 0)$
$3y - 6x = -12$
$2y + 6x = 2$

5. $(-4, -1)$
$2y - x = 2$
$y - 3x = 2$

Use the graph to identify the solution of the linear system it represents.

6.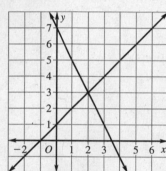

7.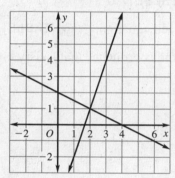

8.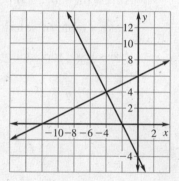

In Exercises 9–17, solve the linear system by graphing.

9. $y = x$
$y = -2x$

10. $y = -x$
$y = 2x - 3$

11. $y = 2x + 1$
$y = -x + 4$

12. $y = 4x + 2$
$y = 4x + 5$

13. $3y = -9x + 3$
$2y + 6x = 2$

14. $3y + 2x = 4$
$2y - 6x = -12$

15. $y = 5$
$2y - 2x = 4$

16. $2y - x = 4$
$6y - 12 = 3x$

17. $3y - 6x = 6$
$2y = 4x - 6$

18. You play racquetball at a community club. You have two options for paying for court time.

Option A: You pay $12 for court time each time you play.

Option B: You buy a club membership for $120 and then pay $2 each time you play.

a. Write and solve a linear system to determine the number of visits for which the cost would be the same for each option.

b. For what numbers of visits is option B less expensive?

Name _____ Date _____

Practice B
For use with pages 431–435

Tell whether the ordered pair is a solution of the linear system.

1. $(-2, 1)$

$y - 2x = 5$

$2y + 2x = -2$

2. $(2, 4)$

$2y - x = 3$

$y + 4 = 4x$

3. $(3, -2)$

$2y - \frac{2}{3}x = 6$

$\frac{1}{2}y + 2x = 5$

In Exercises 4–12, solve the linear system by graphing.

4. $y = 4$

$x = 3$

5. $y = -3x - 7$

$15x + 5y = -35$

6. $y = 3x + 3$

$y + 3x = 3$

7. $4x + 5y = 10$

$3x - 3y = 21$

8. $6x - 10y = -10$

$5y = 3x + 5$

9. $6x - 4y = -2$

$-3x + 5y = -11$

10. $y = \frac{3}{4}x + 8$

$4y = 3x + 4$

11. $\frac{3}{2}y - x = 3$

$3y = 2x - 6$

12. $\frac{4}{5}x - \frac{2}{5}y = 2$

$2y = -3x + 4$

13. You are planning a family reunion. You have two options to pay for food and the banquet hall. For the first option, you can pay $240 for the use of the banquet hall and pay a caterer $8 per person for food. For the second option, the banquet hall will also furnish the food for $600.

a. Write a system of equations describing the total cost of the family reunion.

b. Solve the system of equations. After how many family members are the two options equal?

c. When does the first option have the lower total cost? When does the second option have the lower cost?

14. A car enters a highway at point A and travels west at a constant speed of 55 miles per hour. One hour later, another car enters the highway at point A, and travels west at a constant speed of 65 miles per hour. How long does it take the second car to overtake the first?

15. The graphs of the three equations below form a triangle. Find the coordinates of the triangle's vertices.

$4x + y = 1$

$2x - y = 5$

$3y + 3x = 12$

Name _____ Date _____

Practice C

For use with pages 431–435

Tell whether the ordered pair is a solution of the linear system.

1. $\left(9, \dfrac{1}{5}\right)$

$\dfrac{1}{3}x + 5y = 4$

$\dfrac{4}{3}x + 20y = 16$

2. $(10, 30)$

$\dfrac{16}{5}x - \dfrac{7}{10}y = 11$

$5x + \dfrac{5}{6}y = 75$

3. $(22, 9)$

$\dfrac{4}{11}x + \dfrac{1}{3}y = 11$

$\dfrac{2}{7}x - \dfrac{7}{9}y = -\dfrac{1}{2}$

In Exercises 4–12, solve the linear system by graphing.

4. $3x + 14y = -46$
$5x + 5y = 15$

5. $21y = 9x + 2$
$7y = 3x + 2$

6. $y - 6x = 105$
$3y - 10x = 315 + 8x$

7. $x + y = 18$
$3y - 4x = 19$

8. $-5y + 12x = 150$
$10y - 8x = 100$

9. $5x - 7y = 24$
$-11x + 9y = -8$

10. $4x + 3y = -17$
$2y = -\dfrac{8}{3}x + 5$

11. $5y + 4x = 80$
$2y = -\dfrac{8}{5}x + 32$

12. $\dfrac{3}{7}y + \dfrac{2}{3}x = -10$
$y = \dfrac{11}{2}x + 19$

In Exercises 13–15, tell how many values of m and how many values of b in the system below result in the given number of solutions for the system.

$$y = -7x + 23$$
$$y = mx + b$$

13. Exactly one

14. None

15. Infinitely many

16. A recycler pays $.71 per pound of aluminum and $1.02 per pound of copper. One day the recycler pays $496.80 for a total of 560 pounds of scrap aluminum and copper.

  a. Write a system of equations to represent the situation.

  b. Use a graphing calculator to solve the system of equations. How many pounds of copper and how many pounds of aluminum did the recycler buy?

  c. How much would the recycler pay out if there were equal amounts of scrap aluminum and copper totaling 560 pounds?

Lesson 8.8

Name _____ Date _____

Study Guide

For use with pages 431–435

GOAL Graph and solve systems of linear equations.

VOCABULARY

A **system of linear equations,** or simply a *linear system*, consists of two or more linear equations with the same variables.

A **solution of a linear system** in two variables is an ordered pair that is a solution of each equation in the system.

EXAMPLE 1 Solving a System of Linear Equations

Solve the linear system: $y = 4x - 2$ **Equation 1**
$y = 3x + 1$ **Equation 2**

(1) Graph the equations.

(2) Identify the apparent intersection point, (3, 10).

(3) Verify that (3, 10) is the solution of the system by substituting 3 for x and 10 for y in each equation.

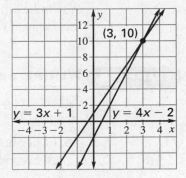

Equation 1	**Equation 2**
$y = 4x - 2$	$y = 3x + 1$
$10 \stackrel{?}{=} 4(3) - 2$	$10 \stackrel{?}{=} 3(3) + 1$
$10 = 10 \checkmark$	$10 = 10 \checkmark$

Answer: The solution is (3, 10).

EXAMPLE 2 Solving a Linear System with No Solution

Solve the linear system: $y = 3x + 5$ **Equation 1**
$y = 3x - 2$ **Equation 2**

Graph the equations. The graphs appear to be parallel lines. You can confirm that the lines are parallel by observing from their equations that they have the same slope, 3, but different y-intercepts, 5 and -2.

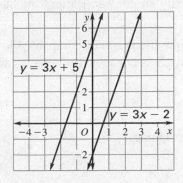

Answer: Because parallel lines do not intersect, the linear system has no solution.

Lesson 8.8

Name _____   Date _____

Study Guide
For use with pages 431–435

EXAMPLE 3 **Solving a Linear System with Many Solutions**

Solve the linear system: $2y + 14x = -6$ **Equation 1**

$3 + y = -7x$ **Equation 2**

Write each equation in slope-intercept form.

Equation 1	**Equation 2**
$2y + 14x = -6$	$3 + y = -7x$
$2y = -14x - 6$	$y = -7x - 3$
$y = -7x - 3$	

$y = -7x - 3$

The slope-intercept forms of equations 1 and 2 are identical, so the graphs of the equations are the same line.

Answer: Because the graphs have infinitely many points of intersection, the system has infinitely many solutions. Any point on the line $y = -7x - 3$ represents a solution.

EXAMPLE 4 **Writing and Solving a Linear System**

Cable company A charges $50 per month, plus an initial set-up fee of $80. Cable company B charges $40 per month, plus an initial fee of $150.

a. After how many months are the total charges of the cable companies the same?

b. When is company B's cable a better deal?

Solution

a. Let y be the charges of each company after x months.

Company A: $y = 50x + 80$

Company B: $y = 40x + 150$

Use a graphing calculator to graph the equations. Trace along one of the graphs until the cursor is on the point of intersection. This point is $(7, 430)$.

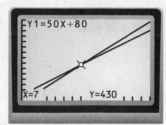

Answer: The total charges of each company are the same after 7 months, when each company charges $430.

b. The graph for company B's charges is below the graph of company A's charges when $x > 7$, so cable with company B is cheaper if you have cable for more than 7 months.

Exercises for Examples 1–4

In Exercises 1–3, solve the linear system by graphing.

1. $2y = 14x + 18$
$3y - 21x = 27$

2. $y + 6 = 4x$
$y - 8x = -2$

3. $5y = -45x + 5$
$27x + 3y = -9$

4. Festival A charges $6 admission plus $1 per ride. Festival B charges $2 admission plus $2 per ride. When is festival A a better deal?

Name _____ Date _____

Challenge Practice

For use with pages 431–435

Solve the linear system by graphing.

1. $5x + 2y = 6$
$3y + 7x = 8$

2. $2y - 3x + 5 = 0$
$3y + 4x - 18 = 0$

3. $2y + 5 = x$
$2y - 1 = 7x$

4. $y = 2x$
$y = -3x + 5$
$y = 2$

5. $4x + 3y - 24 = 0$
$\frac{2}{3}x + \frac{1}{2}y = 4$
$y + \frac{4}{3}x = 8$

6. $y = 3x - 2$
$y = -4x + 12$
$y = -3x + 9$

7. Write a linear equation so that the linear system formed by your equation
and the equation $y = 2x - 3$ has the given solution.

   a. one solution at $(3, 3)$

   b. no solution

   c. many solutions

8. Use a system of equations to find the dimensions of a rectangle so that its
length is three times its width and its perimeter is 56 units.

Teacher's Name _____ Class _____ Room _____ Date _____

Lesson Plan

2-day lesson (See *Pacing and Assignment Guide*, TE page 382A)

For use with pages 436–441

GOAL **Graph inequalities in two variables.**

State/Local Objectives

_____

_____

✓ Check the items you wish to use for this lesson.

STARTING OPTIONS

_____ Homework Check (8.8): TE page 434; Answer Transparencies

_____ Homework Quiz (8.8): TE page 435; Transparencies

_____ Warm-Up: Transparencies

TEACHING OPTIONS

_____ Notetaking Guide

_____ Examples: Day 1: 2, 4, SE pages 437–438; Day 2: 1, 3, SE pages 436–437

_____ Extra Examples: TE pages 437–438

_____ Checkpoint Exercises: Day 1: 1, 4, SE pages 437–438; Day 2: 2–3, SE page 437

_____ Concept Check: TE page 438

_____ Guided Practice Exercises: Day 1: 1–2, 7–8, 11, SE page 439; Day 2: 3–6, 9–10, SE page 439

_____ Technology Activity: CRB page 76

APPLY/HOMEWORK

Homework Assignment

_____ Basic: Day 1: pp. 439–441 Exs. 12–14, 20–27, 38–43
Day 2: pp. 439–441 Exs. 16–19, 28–33, 44–48

_____ Average: Day 1: pp. 439–441 Exs. 12–15, 24–27, 38–43
Day 2: pp. 439–441 Exs. 16–23, 28–34, 44–48

_____ Advanced: Day 1: pp. 439–441 Exs. 12–15, 24–27, 40–46
Day 2: pp. 439–441 Exs. 18–23, 30–37*, 47, 48

Reteaching the Lesson

_____ Practice: CRB pages 77–79 (Level A, Level B, Level C); Practice Workbook

_____ Study Guide: CRB pages 80–81; Spanish Study Guide

Extending the Lesson

_____ Real-World Problem Solving: CRB page 82

_____ Challenge: SE page 441; CRB page 83

ASSESSMENT OPTIONS

_____ Daily Quiz (8.9): TE page 441 or Transparencies

_____ Standardized Test Practice: SE page 441

_____ Quiz (8.6–8.9): Assessment Book page 97

Notes _____

LESSON
8.9
Lesson Plan for Block Scheduling
1-block lesson (See *Pacing and Assignment Guide*, TE page 382A)

For use with pages 436–441

GOAL Graph inequalities in two variables.

State/Local Objectives _____

_____

✓ **Check the items you wish to use for this lesson.**

Chapter Pacing Guide	
Day	**Lesson**
1	81.; 8.2
2	8.2 (end); 8.3
3	8.4; 8.5
4	8.6
5	8.7; 8.8
6	**8.9**
7	Ch. 8 Review and Projects

STARTING OPTIONS
____ Homework Check (8.8): TE page 434; Answer Transparencies

____ Homework Quiz (8.8): TE page 435; Transparencies

____ Warm-Up: Transparencies

TEACHING OPTIONS
____ Notetaking Guide

____ Examples: 1–4, SE pages 436–438

____ Extra Examples: TE pages 437–438

____ Checkpoint Exercises: 1–4, SE pages 437–438

____ Concept Check: TE page 438

____ Guided Practice Exercises: 1–11, SE page 439

____ Technology Activity: CRB page 76

APPLY/HOMEWORK

Homework Assignment

____ Block Schedule: pp. 439–441 Exs. 12–34, 38–48

Reteaching the Lesson

____ Practice: CRB pages 77–79 (Level A, Level B, Level C); Practice Workbook

____ Study Guide: CRB pages 80–81; Spanish Study Guide

Extending the Lesson

____ Real-World Problem Solving: CRB page 82

____ Challenge: SE page 441; CRB page 83

ASSESSMENT OPTIONS
____ Daily Quiz (8.9): TE page 441 or Transparencies

____ Standardized Test Practice: SE page 441

____ Quiz (8.6–8.9): Assessment Book page 97

Notes _____

_____

_____

Name _____ Date _____

Technology Activity

For use with pages 436–441

GOAL Use a calculator to graph linear inequalities.

EXAMPLE Use a graphing calculator to graph $x + y > 2$.

Solution

1 Isolate y on the left side of the inequality.

$y > -x + 2$

2 Enter the inequality.

Keystrokes: $+$ 2

3 Highlight the symbol to the left of Y1. Then press **ENTER** twice to select the icon that will shade above the line.

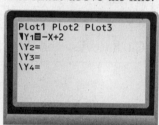

4 Press **GRAPH** to graph the inequality.

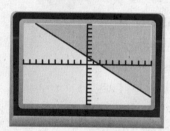

Technology Tip

In this activity, you shaded the half-plane above the line $y = -x + 2$ because of the greater than symbol. If the equation was $y < -x + 2$, you would select the icon to shade below the line.

DRAW CONCLUSIONS Sketch the graph of the inequality in a coordinate plane. Then use a graphing calculator to check your graph.

1. $y > x + 4$ **2.** $y < -2x + 1$ **3.** $y < x - 6$

4. $y > -3x - 5$ **5.** $x + 2y < 6$ **6.** $3x - 2y < 12$

7. You have 12 cups of fresh pumpkin for baking. You need $1\frac{1}{2}$ cups for a pumpkin pie and $\frac{3}{4}$ cup for a loaf of pumpkin bread. Write an inequality describing the possible numbers of pies and loaves of bread that you can make. Then graph the inequality using a graphing calculator.

Lesson 8.9

Name _____ Date _____

Practice A

For use with pages 436–441

Describe and correct the error in the graph of the given inequality.

1. $y < x + 3$

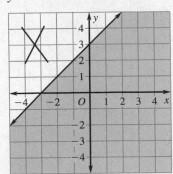

2. $y \geq 4x - 5$

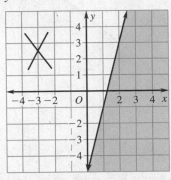

Tell whether the ordered pair is a solution of $3x + y < 2$.

3. $(1, 2)$ **4.** $(-1, -2)$ **5.** $(0, -4)$ **6.** $(-1, 8)$

Graph the inequality in a coordinate plane.

7. $y > 5x - 2$ **8.** $x < 3$ **9.** $y \geq -2$

10. $y \leq x - 5$ **11.** $y < -2x$ **12.** $y \geq -x + 4$

13. $x + 2y < -6$ **14.** $3y - 2x > 12$ **15.** $4x + 5y \leq -5$

16. $9x + 8y > 80$ **17.** $2y - 5x \geq 12$ **18.** $7x + 6y \leq -18$

19. You have 10 feet of yarn to use for making bracelets and necklaces. You need 1 foot for each bracelet and 2 feet for each necklace.

    a. Write an inequality describing the possible numbers of bracelets and necklaces that you can make.

    b. Graph the inequality from part (a).

    c. Give three possible combinations of bracelets and necklaces that you can make.

20. You have no more than 15 hours this week to spend reading a book and writing in your journal. Write and graph an inequality describing the possible amounts of time you can spend reading and writing this week.

Write an inequality to represent the graph.

21.

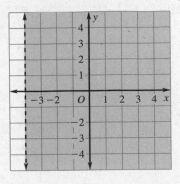

22.

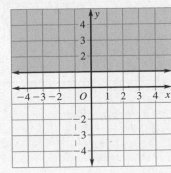

23.

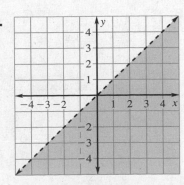

Name _____ Date _____

Practice B

For use with pages 436–441

Describe and correct the error in the graph of the given inequality.

1. $y \le -\dfrac{5}{6}x + 3$

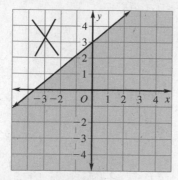

2. $y > -4x$

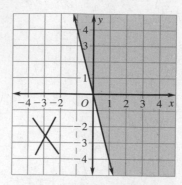

Tell whether the ordered pair is a solution of the inequality.

3. $y \le -9x - 1;\ (-1, -5)$

4. $y < 2x + 3;\ (2, 2)$

5. $x \ge -7;\ (3, -1)$

6. $2x + 3y > 15;\ (1, 4)$

Graph the inequality in a coordinate plane.

7. $-3x - 4y > -36$

8. $7x + 4y < -32$

9. $10y - 3x \le -50$

10. $2x + y \ge 21$

11. $9y - 4x \ge 18$

12. $x + 2y < -11$

13. $y > -1$

14. $x \ge 5$

15. $y \le 3$

16. You have $42 to spend on museum tickets for a group of adults and children. Adult tickets are $3 and child tickets are $1.50.

  a. Write an inequality describing the possible numbers of adult and child tickets that you can buy.

  b. Graph the inequality from part (a).

  c. Give three possible combinations of adult and child tickets that you can buy.

17. You are selling T-shirts and buttons as a fund-raiser for your soccer team. You want to earn at least $136. The T-shirts cost $12 and the buttons cost $4. Write and graph an inequality describing the possible numbers of T-shirts and buttons you could sell.

Write an inequality to represent the graph.

18.

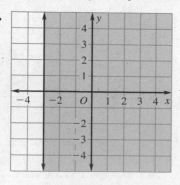

19.

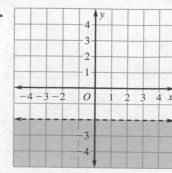

20.

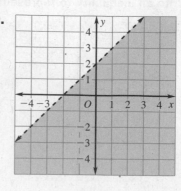

Name _____ Date _____

Practice C

For use with pages 436–441

Tell whether the ordered pair is a solution of the inequality.

1. $y < -6x + 8$; $(5, -25)$

2. $y > 12x - 7$; $(-4, 6)$

3. $x \leq 7$; $(11, -1)$

4. $y \geq -4$; $(3, -2)$

5. $6x + 2y < 18$; $(16, -7)$

6. $4y - x \geq -32$; $(-8, -10)$

Graph the inequality in a coordinate plane.

7. $x \leq -10$

8. $y \geq 12$

9. $y < \frac{1}{3}x - 5$

10. $y > -\frac{3}{5}x - 7$

11. $7y - 5x \leq -7$

12. $8x + 3y \geq 6$

13. $9x + 4y > -12$

14. $y - 5x \leq -12$

15. $3x + y > 24$

16. $x + 6y \leq -72$

17. $x + 8y \geq 88$

18. $5y - 4x < -100$

19. You have two part time jobs. One is at a grocery store, where you earn $6 an hour, and the other is mowing lawns, where you earn $5 an hour. Between the two jobs, you want to earn at least $82 a week. Write and graph an inequality describing the possible hours you can work at each job. Give three possible combinations of hours worked at each job.

Write an inequality to represent the graph.

20.
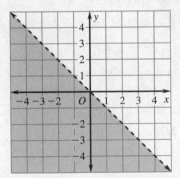

21.
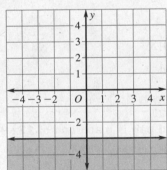

22.
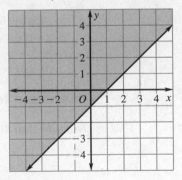

In Exercises 23–25, use the *system* of linear inequalities shown below. A *solution* of the system is an ordered pair that is a solution of each inequality.

$$y < x - 2$$
$$y \geq -\frac{1}{2}x + 1$$

23. Tell whether each ordered pair is a solution of the system.

    a. $(-1, 2)$

    b. $(4, 0)$

    c. $(8, -2)$

24. Graph the inequalities in the system. Draw both graphs in the same coordinate plane, and shade only the portion of the graph that is a solution of both inequalities.

25. Describe the region of the plane that contains the solutions of the system.

LESSON

8.9 Study Guide

For use with pages 436–441

Name _____ Date _____

GOAL Graph inequalities in two variables.

VOCABULARY

A **linear inequality** in two variables, such as $2x - 3y < 6$, is the result of replacing the equal sign in a linear equation with $<$, $\leq$, $>$, or $\geq$.

An ordered pair (x, y) is a **solution of a linear inequality** if substituting the values of x and y into the inequality produces a true statement.

The **graph of a linear inequality** in two variables is the set of points in a coordinate plane that represent all the inequality's solutions. The graph of a line in a coordinate plane divides the plane into two half-planes.

EXAMPLE 1 Checking Solutions of a Linear Inequality

Tell whether the ordered pair is a solution of $5x + 4y \geq 45$.

a. $(1, 10)$ **b.** $(-5, 3)$

Solution

a. Substitute 1 for x and 10 for y.

$$5x + 4y \geq 45$$
$$5(1) + 4(10) \overset{?}{\geq} 45$$
$$45 \geq 45$$

$(1, 10)$ is a solution.

b. Substitute -5 for x and 3 for y.

$$5x + 4y \geq 45$$
$$5(-5) + 4(3) \overset{?}{\geq} 45$$
$$-13 \not\geq 45$$

$(-5, 3)$ is not a solution.

Exercises for Example 1

Tell whether the ordered pair is a solution of $12x + 3y > 100$.

1. $(5, 14)$ **2.** $(3, 21)$ **3.** $(8, 1)$ **4.** $(9, 0)$

EXAMPLE 2 Graphing a Linear Inequality

Graph $y < 3x - 5$.

(1) Draw the boundary line $y = 3x - 5$. The inequality symbol is $<$, so use a dashed line.

(2) Test the point $(0, 0)$ in the inequality.

$$y < 3x - 5$$
$$0 \overset{?}{<} 3(0) - 5$$
$$0 \not< -5$$

(3) Because $(0, 0)$ is not a solution, shade the half-plane that does *not* contain $(0, 0)$.

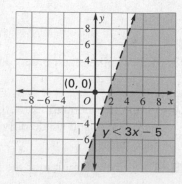

Vertical label (left margin): Lesson 8.9

LESSON
8.9
Continued

Study Guide

For use with pages 436–441

EXAMPLE 3 **Graphing Inequalities with One Variable**

Graph $x \geq -4$ and $y < -6$ in a coordinate plane.

a. Graph $x = -4$ using a solid line. Use $(0, 0)$ as a test point.

$$x \geq -4$$

$$0 \geq -4 \checkmark$$

Shade the half-plane that contains $(0, 0)$.

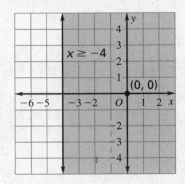

b. Graph $y = -6$ using a dashed line. Use $(0, 0)$ as a test point.

$$y < -6$$

$$0 \not< -6$$

Shade the half-plane that does *not* contain $(0, 0)$.

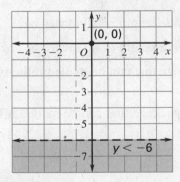

Exercises for Examples 2 and 3

Graph the inequality in a coordinate plane.

5. $y \geq 4$ **6.** $y \leq -4x + 1$ **7.** $5x - y > 2$ **8.** $x < -3$

Name _____ Date _____

Real-World Problem Solving

For use with pages 436-441

Mailing a Package

When mailing a package, the U.S. Postal Service requires that the height of
a package combined with the girth, or the distance around the package, cannot
be greater than 108 inches.

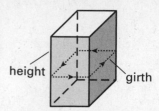

Consider the package below. It has a square base with side length x and height y.

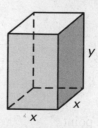

In Exercises 1–6, use the information above.

1. Write an inequality that represents the situation above.

2. Graph the inequality.

3. What heights can a package with a side length of 20 inches be?

4. What side lengths can a package with a height of 40 inches be?

5. What is the longest height of a package that can be mailed if the package
has a side length of 23 inches?

6. What is the longest side length of a package that can be mailed if the
package has a height of 52 inches?

Lesson 8.9

Challenge Practice

For use with pages 436–441

Graph the inequality in a coordinate plane.

1. $5x + 4y \geq 12$ **2.** $3x - 2y < 5$ **3.** $6y - 4x > 8$

Write the inequality that is represented by the graph.

4.

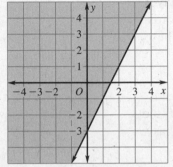

5.

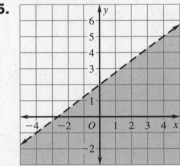

6.

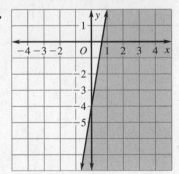

7. Write and graph a linear inequality that has (2, 4) as a solution point, but does not have (−1, 5) and (3, −2) as solution points.

Graph the inequalities in the system of linear inequalities in the same coordinate plane. Then find the area of the region enclosed by the inequalities.

8. $x \geq 0$
$y \geq 0$
$y \leq 4 - x$

9. $y \geq 0$
$y \leq x$
$y \leq 6 - x$

10. $x \geq 0$
$y \geq 0$
$y \leq 4$
$x \leq 6$

8 Chapter Review Games and Activities

For use after Chapter 8

Vocabulary Bingo

Materials: textbook, index cards, tokens

Preparation   On a sheet of paper, each player draws a four-block by four-block grid. Randomly choose and write 16 of the vocabulary words listed below in the blocks. Each player's completed grid should look different from all other player's grids. Next, use your textbook to write the definitions of the vocabulary words listed below on separate index cards. Also, each player will need tokens. These can be made out of paper if needed. The tokens should be the same size or smaller than the blocks from the grid.

To Play   Shuffle the index cards. One player chooses an index card and reads the definition. All players then place a token on their grid, in the block with the matching vocabulary word. Continue picking index cards until one player has covered four blocks in a row.

To Win   The player with four covered blocks in a row, either vertically, horizontally, or diagonally, and has correctly matched each covered vocabulary word in the row with its definition, wins the game.

Vocabulary words to write on grid:

relation	vertical line test	rise
domain	linear equation	run
range	function form	slope-intercept form
input	*x*-intercept	best-fitting line
output	*y*-intercept	function notation
function	slope	half-plane

Real-Life Project: Bake Sale

For use after Chapter 8

Objective Determine how many times you can make certain recipes using a given amount of an ingredient.

Materials pencil, paper, graph paper, access to the Internet or library

Investigation *Getting Going*   You are a member of a world languages club. In order to raise money for a trip, the club is having a bake sale. For your donated goods, you choose to make peanut butter cookies and peanut butter pies.

You buy a 40-ounce jar of peanut butter to use for your recipes. The recipe for peanut butter cookies calls for $\frac{3}{4}$ cup of peanut butter. The recipe for peanut butter pie calls for $\frac{1}{2}$ cup of peanut butter.

Questions

1. How many cups of peanut butter are in the 40-ounce jar? Use the conversion factor 1 cup = 8 ounces.

2. Write an inequality describing the possible numbers of batches of cookies and pies that you can make using the 40-ounce jar of peanut butter. Let x represent the number of batches of cookies. Let y represent the number of pies.

3. Graph your inequality from Question 2. Label your boundary line and identify your test point.

4. What does the y-intercept of the boundary line represent? What does the x-intercept of the boundary line represent?

5. Give three possible combinations of batches of cookies and pies that can be made.

6. Redo Questions 2, 3, and 5 for a 28-ounce jar of peanut butter.

7. You have 10 cups of flour to bake cookies. Using the Internet or your library, choose two cookie recipes that you can use for the bake sale. Then repeat Questions 2–5 for your recipes.

Review and Projects

Teacher's Notes for Bake Sale Project

For use after Chapter 8

Project Goals
- Identify *x*- and *y*-intercepts.
- Check solutions of a linear inequality.
- Write and graph a linear inequality.

Managing the Project

Guiding Students' Work   In Question 6, make sure students remember to find the number of cups in the jar of peanut butter before they write their inequality.

When graphing the linear inequalities, remember to explain to students that possible combinations will only occur in Quadrant I or along the axes. Have students check their solutions when finding possible combinations.

Rubric for Project   **The following rubric can be used to assess student work.**

4 The student correctly writes and graphs the inequality in Questions 2 and 3. The graph of the inequality is clear and neat, with the boundary line and test point labeled. The student gives three possible combinations that can be used. All of the questions for the 28-ounce jar of peanut butter are answered correctly. The graph is neat and labeled correctly. The student finds two cookie recipes and answers all the corresponding questions correctly. All of the calculations are correct. The student's work is organized and neat.

3 The student correctly writes and graphs the inequality in Questions 2 and 3. The graph of the inequality is clear and neat, but the boundary line and test point are not labeled. The student gives three possible combinations that can be used. Most of the questions for the 28-ounce jar of peanut butter are answered correctly. The graph is neat, but is not labeled. The student finds two cookie recipes and answers all the corresponding questions with little difficulty. Most of the calculations are correct. The student's work is neat.

2 The student correctly writes the inequality in Question 2, but its graph has some errors. The student gives three possible combinations but one is incorrect. Some of the questions for the 28-ounce jar of peanut butter are answered incorrectly. The graph is a little sloppy. The student finds two cookie recipes but has some difficulty writing the inequality and answering the corresponding questions. Some of the calculations are incorrect. The student's work is sloppy.

1 The student writes an inequality for Question 2 but it is incorrect. The student cannot give three possible combinations. Most of the questions for the 28-ounce jar of peanut butter are answered incorrectly. The graphs are sloppy. The student finds two cookie recipes but has difficulty writing the inequality and answering the corresponding questions. Most of the calculations are incorrect. The student's work is incomplete or sloppy.

Name _____   Date _____

Cooperative Project: Sight and Distance

For use after Chapter 8

Objective Find a best-fitting line that represents a data set for an experiment.

Materials paper, pencil, cardboard tube (from a roll of paper towels), yardstick, tape measure, tape, scissors

Investigation *Getting Going*   This project is for 3 or more students. The object is to find a relationship between the distance a person stands from a wall and how much they can see looking through a cardboard tube. A standard cardboard tube from a roll of paper towels is about 11 inches long.

Using a tape measure or yardstick, mark off distances from a wall as shown below. Beginning with a distance of 2 feet, mark off distances of 4 feet, 6 feet, 8 feet, 10 feet, and so on.

Once the distances are marked, one member of the group will look through the cardboard tube, one will hold a yardstick vertically on the wall as shown, and the other will record the results. The student that looks through the tube will determine how much of the yardstick they can see from each distance. The student holding the yardstick will mark the height. This will be the viewing height. For example, if you can see the yardstick from 5 inches to 14 inches, the viewing height is 9 inches.

Repeat this procedure for each marked distance. When estimating the viewing height, round the distance to the nearest half inch. Make a table and organize the results.

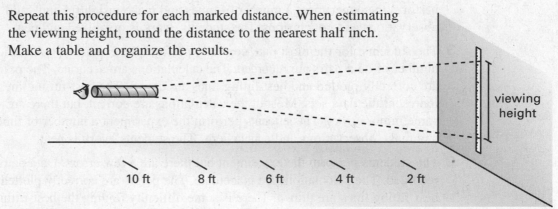

Questions

1. Have each member of your group look through the cardboard tube and record their viewing heights. Record the results for each member of your group and compare the viewing heights. Keep your results in your own table.

2. Use the results in your table to make a scatter plot of the data pairs. Let *x* represent the distance from the wall. Let *y* represent the viewing height on the wall. What relationship(s) do you notice?

3. Draw a best-fitting line for the data points. Find the slope of your line. Then write its equation. Compare your equation with members of your group.

4. Use scissors to cut the cardboard tube down to different lengths. Then repeat the project. What happens to the viewing heights when the cardboard tube is cut shorter? Record your group's observations and conclusions.

Review and Projects

Teacher's Notes for Sight and Distance Project

For use after Chapter 8

Project Goals
- Plot points in a coordinate plane.
- Make a scatter plot.
- Find a best-fitting line.

Managing the Project

Classroom Management   This project can be for more than 3 students. Students can use as many points as they want. Viewing heights can be determined for odd number distances as well.

Each student should keep their calculations in their own table.

Alternative Approach   To shorten the experiment, only one student from each group should look through the cardboard tube. The group can record the viewing heights and find one best-fitting line.

Rubric for Project

The following rubric can be used to assess student work.

4 The students correctly perform the experiment. A sufficient number of data pairs are recorded. The calculations are accurate. The points are correctly plotted and best-fitting lines are drawn. The best-fitting lines correctly represent the data sets. All of the calculations are correct. The students perform the experiment a number of times and make insightful and valid observations and conclusions. The students' work is organized and neat.

3 The students, for the most part, correctly perform the experiment. A sufficient number of data pairs are recorded. The calculations are accurate. The points are correctly plotted and best-fitting lines are drawn. The best-fitting lines represent the data sets. Most of the calculations are correct, but there are some minor errors. The students perform the experiment a number of times and make observations and conclusions. The students' work is neat.

2 The students perform the experiment but there are a few errors. Data pairs are recorded. The calculations are inaccurate. The points are correctly plotted and best-fitting lines are drawn. There is some difficulty finding the best-fitting lines. Some of the calculations are incorrect. The students perform the experiment a number of times but their observations and conclusions are unclear. The students' work is a little sloppy.

1 The students do not correctly perform the experiment or no attempt is made. Data pairs are not accurately recorded. The points are plotted and best-fitting lines are drawn, but there are some errors. There is difficulty finding the equations for the best-fitting lines. Many of the calculations are incorrect. The students do not correctly repeat the experiment and no observations or conclusions are given. The students' work is incomplete or sloppy.

Review and Projects

Independent Extra Credit Project: Consumer Spending

For use after Chapter 8

Objective **Find a best-fitting line that represents a data set that shows the amount of money spent per person for television costs.**

Materials pencil, paper, graph paper, ruler, access to the Internet

Investigation *Getting Going* The tables below show the amount of money spent per person in the United States for television costs including cable and satellite services for the years 1990–2002.

Years since 1990, x	0	1	2	3	4	5	6
Cost per person, y	$88	$94	$101	$108	$108	$122	$139

Years since 1990, x	7	8	9	10	11	12
Cost per person, y	$153	$166	$180	$193	$207	$223

Questions

1. Is the relation a function? Explain your reasoning.

2. Use graph paper to make a scatter plot of the data pairs. Do the points lie on a line?

3. Draw a line that appears to best fit the data points. Find the slope and y-intercept of your line. Then write its equation.

4. Make a table of solutions for your equation for the years 1990–2002.

5. Use the graph to estimate the amount spent per person in 2005. Then use your equation to check your estimate.

6. Estimate the year when the amount spent per person will reach $275.

7. Estimate the year when the amount spent per person will reach $300.

8. Using your equation, what conclusion(s) can you make about the amount spent per person for television costs in the United States?

9. Use the Internet or a statistical abstract to find the amount spent per person in the United States for Internet access. Research information for several years. Make a scatter plot of the data pairs. Then draw a line that appears to best fit the data points. Write an equation of your line.

Review and Projects

Teacher's Notes for Consumer Spending Project

For use after Chapter 8

Project Goals
- Plot points in a coordinate plane.
- Make a scatter plot.
- Find a best-fitting line.

Managing the Project

Guiding Students' Work   You can perhaps show the class examples given by students for best-fitting lines and how they are similar or different, and why each line can be used to represent the data points.

Using their equation for the best-fitting line, students can find their estimates either graphically, or by substituting for y and solving for x. Remind students that to find the year, their solution is added to the year 1990.

Instruct students that when researching Internet costs, to find information for at least 7 years or more so that their scatter plot and best-fitting line represents a range of years. A statistical abstract is the easiest resource to use to research this information.

Rubric for Project

The following rubric can be used to assess student work.

4 The student correctly plots all the points for the scatter plot. The student's best-fitting line is drawn and is reasonable. The student finds an equation for his/her line. The student's estimates are calculated and are reasonable. Conclusions are clear and insightful. All of the calculations are correct with his/her work shown. The student researches information on consumer spending for Internet access, makes a scatter plot, and finds a best-fitting line. The student's work is organized, clear, and neat.

3 The student correctly plots all the points for the scatter plot. The student's best-fitting line is drawn. The student finds an equation for his/her line with little difficulty. The student's estimates are calculated and are reasonable. Most of the calculations are correct. Conclusions are given. The student researches information on consumer spending for Internet access and makes a scatter plot. The student has a little difficulty finding the equation for his/her line. The student's work is neat.

2 The student plots all the points for the scatter plot with little difficulty. The student's best-fitting line is drawn, but he/she has difficulty finding an equation for his/her line. The student's estimates are somewhat reasonable but there are some minor errors. Some of the calculations are incorrect. The student researches information on consumer spending for Internet access, but he/she has difficulty finding information, making a scatter plot, and finding a best-fit line. The student's work is sloppy.

1 The student plots the points for the scatter plot, but it is sloppy. The student's best-fitting line is incorrect or not drawn. Most of the calculations are incorrect. The student has difficulty researching consumer spending for Internet access, or no attempt is made. The student's work is incomplete or sloppy.

Review and Projects

Cumulative Practice

For use after Chapter 8

Evaluate the expression when $a = 7$. **(Lesson 1.1)**

1. $a + 7$ **2.** $13 - a$ **3.** $\dfrac{21}{a}$ **4.** $12a$

5. A hockey player's shooting percentage is defined by the verbal model below. In the 2002-2003 National Hockey League season, Mario Lemieux had 235 shots on goal and a shooting percentage of 11.91%. How many goals did Lemieux have? Round your answer to the nearest goal. (Lesson 2.7)

$$\text{shooting percentage} = \frac{\text{goals}}{\text{shots on goal}}$$

Write an inequality represented by the graph. (Lesson 3.4)

6.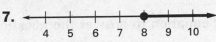

7.

8.

9.

Write the fraction in simplest form. (Lesson 4.3)

10. $\dfrac{9}{30}$ **11.** $\dfrac{20}{45}$ **12.** $\dfrac{21}{35}$ **13.** $\dfrac{84}{102}$

Write the expression without using a fraction bar. (Lesson 4.6)

14. $\dfrac{1}{8}$ **15.** $\dfrac{1}{121}$ **16.** $\dfrac{7}{p^3}$ **17.** $\dfrac{5x^2}{y^4}$

Write the decimal as a fraction or mixed number. (Lesson 5.1)

18. 0.56 **19.** 1.38 **20.** 0.071 **21.** -4.042

Simplify the expression. (Lessons 5.2, 5.3)

22. $\dfrac{4x}{9} + \dfrac{2x}{9}$ **23.** $\dfrac{-10}{13y} - \dfrac{2}{13y}$

24. $\dfrac{5k}{6} + \dfrac{2k}{5}$ **25.** $\dfrac{7m}{10} - \dfrac{m}{4}$

Find the value of x. (Lesson 6.3)

26. $\dfrac{42}{6} = \dfrac{x + 11}{5}$ **27.** $\dfrac{20}{x + 10} = \dfrac{16}{36}$ **28.** $\dfrac{40}{x - 5} = \dfrac{80}{24}$

Review and Projects

Cumulative Practice

For use after Chapter 8

Name the corresponding angles and the corresponding sides of the figures. (Lesson 6.4)

29. $ABCD \cong WXYZ$

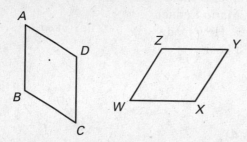

30. $RST \sim JKL$

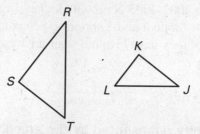

Use the percent equation to answer the question. (Lesson 7.4)

31. What number is 35% of 60?

32. What percent of 85 is 34?

33. 18 is 36% of what number?

34. What percent of 45 is 9?

35. On a trip to New York City, you eat at a local restaurant. The food bill for your meal was $26. The sales tax is 8.25% and you leave a 15% tip. What is the total cost of your meal? (Lesson 7.6)

Identify the domain and range of the relation. (Lesson 8.1)

36. $(-3, 4), (-1, 2), (1, 0), (3, -2), (5, -4)$

37. $(5.2, 0.1), (2.4, 3.8), (7.9, -2.6), (-0.2, 4.7)$

Find the intercepts of the equation's graph. Then graph the equation. (Lesson 8.3)

38. $2x + y = 4$

39. $6x - 9y = 18$

40. $3y = 4x - 12$

41. $x - y = 5$

Write an equation of the line through the given points. (Lesson 8.6)

42. $(1, 4), (0, -1)$

43. $(4, 6), (0, 2)$

44. $(0, -4), (3, -2)$

Answers

Lesson 8.1

Practice A

1. domain **2.** range

3. domain: 0, 1, 2, 3, 4; range: 2, 4, 6, 8, 10

4. domain: 3, 4, 5; range: 4, 5, 6, 7, 8

5. domain: 0, 2, 3; range: 5, 6

6. domain: 1, 4, 7, 10, 13;
range: −5, −4, −3, −2, −1

7. 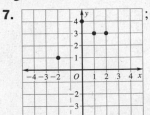 ;

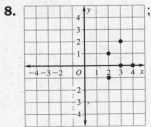 **Input** **Output**; yes; every input is paired with exactly one output.

8. ;

Input **Output**; no; *Sample answer:* the input 2 is paired with two outputs, −1 and 1.

9. 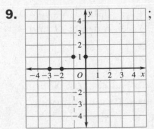 ;

Input **Output**; yes; every input is paired with exactly one output.

10. 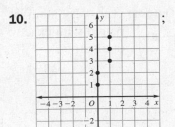 ;

Input **Output**; no; *Sample answer:* the input 0 is paired with two outputs, 1 and 2.

11. 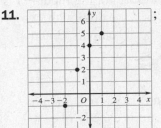 ;

Input **Output**; yes; every input is paired with exactly one output.

12. 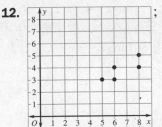 ;

Input **Output**; no; *Sample answer:* the input 6 is paired with two outputs, 3 and 4.

13. 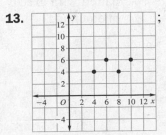 ;

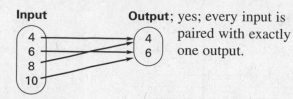

Input **Output**; yes; every input is paired with exactly one output.

4 → 4
6 → 6
8
10

14.

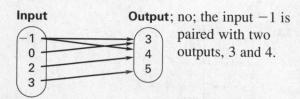

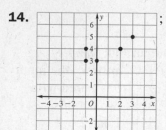

Input **Output**; no; the input −1 is paired with two outputs, 3 and 4.

−1 → 3
0 → 4
2 → 5
3

15. no **16.** yes **17.** no **18.** Yes. *Sample answer:* For each day, there is exactly one precipitation amount. **19.** Yes. *Sample answer:* For every input (number of aluminum cans) there is one output (money received).

Practice B

1. domain: 5, 6, 7, 8; range: 0, 6, 8, 10

2. domain: −6, −3, 4, 7; range: 4, 0, 2, 3, 9

3. domain: −4, −1, 3, 4; range: −5, −4, −3, 2, 0 **4.** domain: 0, 2, 4, 8; range: −3, −1, 1, 3

5.

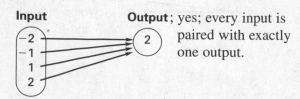

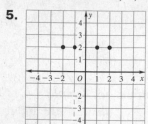

Input **Output**; yes; every input is paired with exactly one output.

−2 → 2
−1
1
2

6.

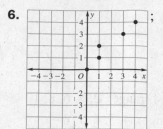

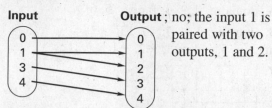

Input **Output**; no; the input 1 is paired with two outputs, 1 and 2.

0 → 0
1 → 1
3 → 2
4 → 3
4

7.

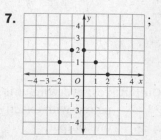

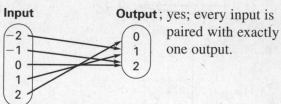

Input **Output**; yes; every input is paired with exactly one output.

−2 → 0
−1 → 1
0 → 2
1
2

8.

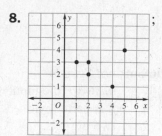

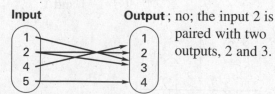

Input **Output**; no; the input 2 is paired with two outputs, 2 and 3.

1 → 1
2 → 2
4 → 3
5 → 4

9. yes **10.** no **11.** yes **12.** Yes. *Sample answer:* Each child pays exactly one amount.

13. a. domain: 55, 75, 102, 110; range: 1002, 1018, 1023, 1250, 1450

Lesson 8.1 *continued*

b.
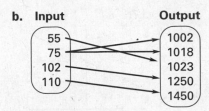

c. No; the input 75 is paired with two outputs, 1002 and 1018.

Practice C

1. domain: 1, 2, 4, 5, 7; range: 0, 4

2. domain: $-9, -2, 2, 5, 7$; range: $-2, -1, 6, 7$

3. domain: $-6, -1, 1, 4, 8$; range: $-4, -1, 2, 6$

4. domain: $-3, -2, 0, 2, 3$; range: $-9, -3, -1, 3, 6$

5.
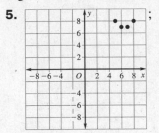
Input ⟶ Output; yes; every input is paired with exactly one output.

6.
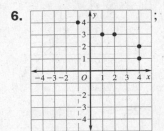
Input ⟶ Output; no; the input 4 is paired with two outputs, 1 and 2.

7. $(-5, 4), (-4, 2), (-3, 2), (-2, 4), (-1, 1)$; yes; each first coordinate is paired with exactly one second coordinate. **8.** $(-5, 3), (-4, 2), (-3, 4), (-2, -2), (-1, 1), (0, 1), (1, 2)$; yes; each first coordinate is paired with exactly one second coordinate.

9. $(-5, -1), (-4, 4), (-3, 2), (-3, -2), (-2, 1), (0, 0), (1, 1), (2, 2)$; no; the first coordinate -3 is paired with both 2 and -2.

10. Yes. *Sample answer:* For each number of quarts, there is exactly one amount of earnings.

11. a. No. *Sample answer:* Working the same number of hours on different shifts will produce different earnings. **b.** No. *Sample answer:* The same earnings are possible for working different numbers of hours if it is done on different shifts. **c.** *Sample answer:* By dividing the data into three sets of ordered pairs (hours worked, earnings) that represent the amounts made during first, second, and third shifts, the ordered pairs in each set represent functions. **12.** *Sample answer:* Sometimes you will travel different distances in the same amount of time. If you always traveled at a steady speed, the situation would be a function.

Study Guide

1. Domain: $-1, 0, 2, 3$ Range: $-3, 0, 4, 7, 8$;

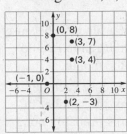 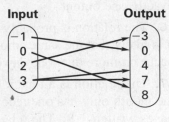

2. Domain: $-4, -1, 5$; Range: $-8, -6, -2$;

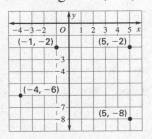 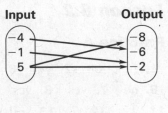

3. The relation is not a function because the input -10 has more than one output, 4 and 5.

4. The relation is a function because every input is paired with exactly one output.

Challenge Practice

1. Domain: $-3, -2, -1, 0, 1, 2, 3$; Range: $-1, 1$; function

Lesson 8.1 *continued*

2. Domain: $-3, -2, -1, 0, 1, 2, 3$; Range: $-1,$ $0, 1, 2$; function   **3.** Domain: $-1, 0, 1, 2$; Range: $-3, -2, -1, 0, 1, 2, 3$; not a function

4.

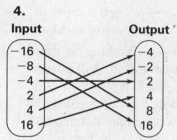
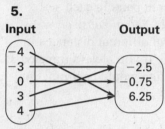

The relation is a function because every input has exactly one output.

5.

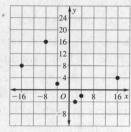
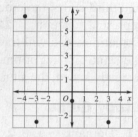

The relation is a function because every input has exactly one output.

6. The relation is probably not a function because it is likely that the same amount of rainfall could occur on more than one day of the month.

7. The relation is a function because each day of the month only has one record of the number of pages written.   **8.** The relation is probably not a function because you could run the same distance in two different times.

Lesson 8.2

Practice A

1. function   **2.** yes   **3.** no   **4.** no   **5.** yes

6. no   **7.** yes   **8.** yes   **9.** no   **10.** 20

11. 12   **12.** 7   **13.** -10

14.

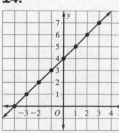

15.

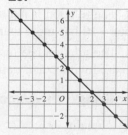

16.

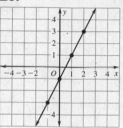

17.

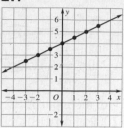

18.

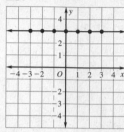

19.

20.

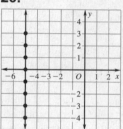

21.

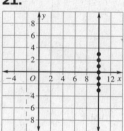

22.

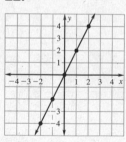

23. $y = -x + 8$;

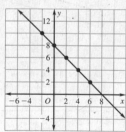

24. $y = 5x + 1$;

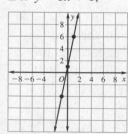

25. $y = 2x + 6$;

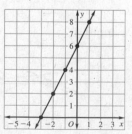

Lesson 8.2 *continued*

26. $y = -x + 2$

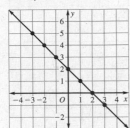

27. $y = -\frac{3}{2}x + 1$;

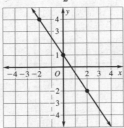

28. $y = x - \frac{7}{3}$;

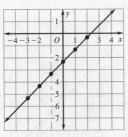

29. 6.46 km **30.** 88 mi

Practice B

1. no **2.** yes **3.** yes **4.** no **5.** 160

6. 306 **7.** 60 **8.** 348

9.

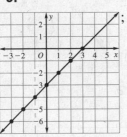

function

10.

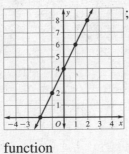

function

11.

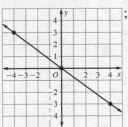

function

12.

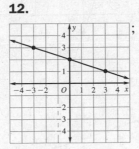

function

13.

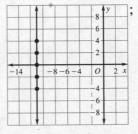

not a function

14.

function

15.

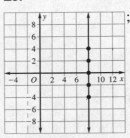

not a function

16.

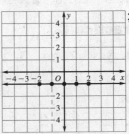

function

17.

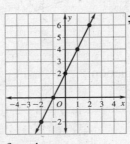

function

18. $y = 7x$;

19. $y = -15x + 20$;

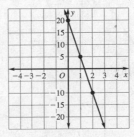

20. $y = -6x + 12$;

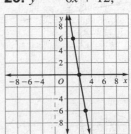

21. $y = \frac{1}{2}x + 2$;

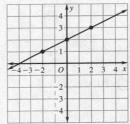

22. $y = \frac{3}{2}x - 3$;

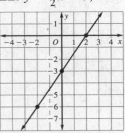

23. $y = \frac{1}{3}x + 2$;

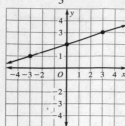

24. About 2800 pounds **25.** $32,400 **26.** −2

27. 3 **28.** 6 **29.** 5

Practice C

1. yes **2.** no **3.** yes **4.** no **5.** −33

6. 34 **7.** 22 **8.** 65

9.

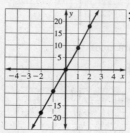

function

10.

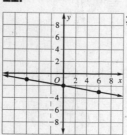

function

11.

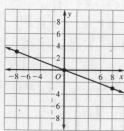

function

12.

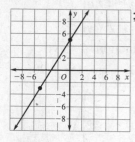

function

13.

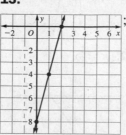

function

14.

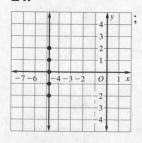

not a function

15.

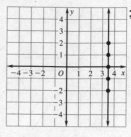

not a function

16.

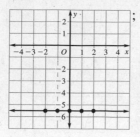

function

17.

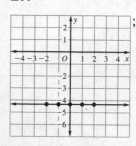

function

18. $y = -x + 1$;

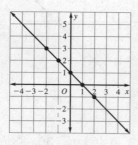

19. $y = x - 6$;

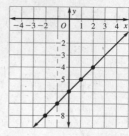

20. $y = \frac{7}{5}x - 2$;

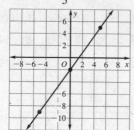

21. $y = -\frac{11}{4}x - 3$;

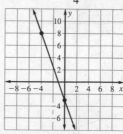

22. $y = \frac{9}{8}x - \frac{1}{2}$;

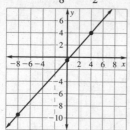

23. $y = -\frac{1}{3}x + 2$;

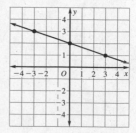

Lesson 8.2 continued

24. *Sample answer:* Use the equation to find and plot the coordinates of three points that are not all in one line. **25.** 10,080,000 cubic meters

26. a. $y = 2204.6x$ **b.** 1322.76 pounds

Study Guide

1. yes **2.** no **3.** no **4.** yes

5.
x	−2	−1	0	1	2
y	21	18	15	12	9
; 5

6.
x	−8	−4	0	4	8
y	−6	−1	4	9	14
; $-3\frac{1}{5}$

7.
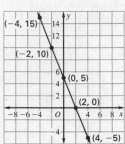

8.
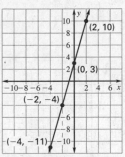

9.
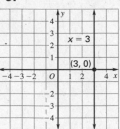

10.
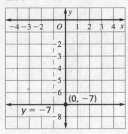

Challenge Practice

1. $y = -\frac{4}{3}x - 8$;
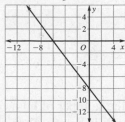

2. $y = 20x - 34$;
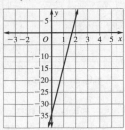

3. $y = -\frac{15}{16}x - \frac{5}{4}$;
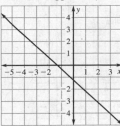

4. $y = -0.7x + 0.41$;
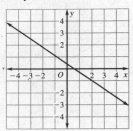

5. −9 **6.** −3 **7.** −9 **8.** 5

9. The graphs intersect at $\left(\frac{1}{2}, \frac{5}{2}\right)$.
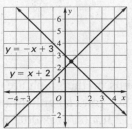

10. The graphs are parallel lines.
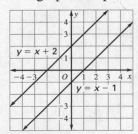

Lesson 8.3

Practice A

1. x-intercept: 2; y-intercept: 3

2. x-intercept: −4; y-intercept: 1

3. x-intercept: 0; y-intercept: 0

4.
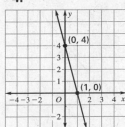

5.
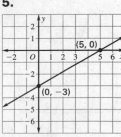

Lesson 8.3 *continued*

6.

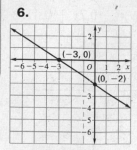

7.

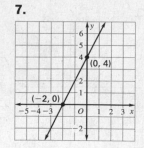

8.

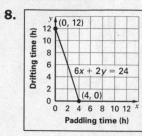

Sample answer: Three points on the graph are (0, 12), (2, 6), and (4, 0). So, you can either not paddle at all and drift for 12 hours, or paddle for 2 hours and drift for 6 hours, or paddle for 4 hours and not drift at all.

9.

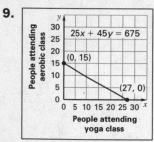

Sample answer: Three points on the graph are (0, 15), (9, 10), and (27, 0). So, either no one joined the yoga class and 15 people joined the aerobic class, or 9 people joined the yoga class and 10 people joined the aerobic class, or 27 people joined the yoga class and no one joined the aerobic class.

10.

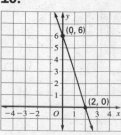

11.

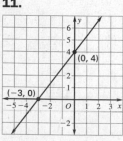

12.

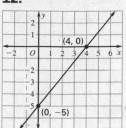

13.

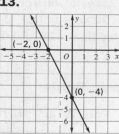

14.

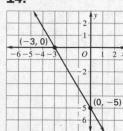

15.

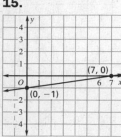

16.

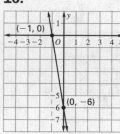

17.

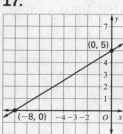

18.

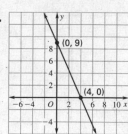

19. a. x-intercept: $7\frac{1}{3}$; y-intercept: 22;

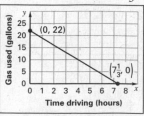

b. The x-intercept represents the number of hours you can drive until your fuel tank is empty. The y-intercept represents the amount of gasoline in gallons in the gas tank of the SUV. **c.** $3\frac{2}{3}$ hours

Lesson 8.3 continued

Practice B

1. *x*-intercept: 4; *y*-intercept: 3
2. *x*-intercept: −2; *y*-intercept: 1
3. *x*-intercept: 1; *y*-intercept: −2

4.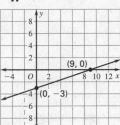

5.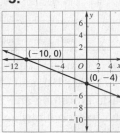

6.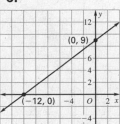

7.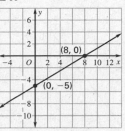

8.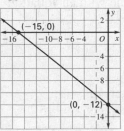

9.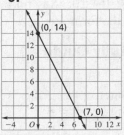

10.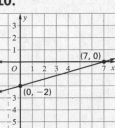

11.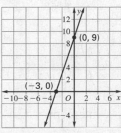

12.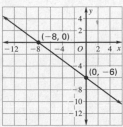

13.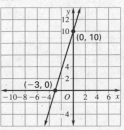

14.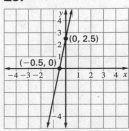

15.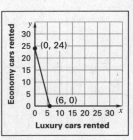

16. Let *x* be the amount of potato salad and *y* be the amount of pasta salad (both in pounds). Then $1.25x + 2.5y = 20$;

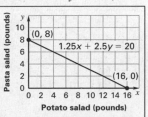

17. *x*-intercept: 6;
y-intercept: 24;
When 4 luxury cars
have been rented,
8 economy cars
have been rented.

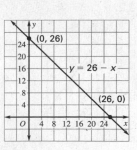

18. a. $52 = 2x + 2y$

b. *x*-intercept: 26;
y-intercept: 26

c. *Sample answer:* (4, 22), (10, 16), and (25, 1).

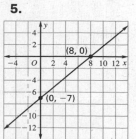

Practice C

1. *x*-intercept: −2; *y*-intercept: −4
2. *x*-intercept: −3; *y*-intercept: 2
3. *x*-intercept: 3; *y*-intercept: −3

4.

5.

Lesson 8.3 *continued*

6.

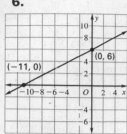

7.

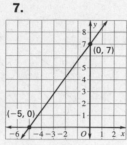

8.

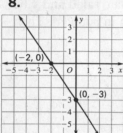

9.

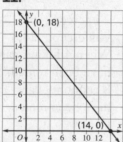

10.

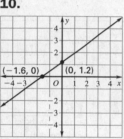

11.

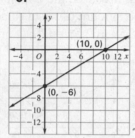

12.

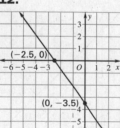

13.

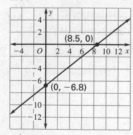

14.

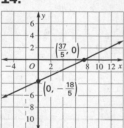

15.

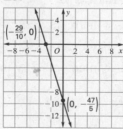

16.

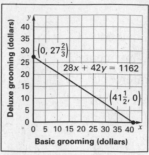

Sample answer: Three points on the graph are (1, 27), (13, 19), and (40, 1). So, either there was 1 basic grooming and 27 deluxe groomings, or 13 basic groomings and 19 deluxe groomings, or 40 basic groomings and 1 deluxe grooming.

17. a. $18 = 2x + y$

b. x-intercept: 9;
   y-intercept: 18;

c. *Sample answer:* Any of these ordered pairs would form a triangle with the given conditions.
(5, 8); (6, 6); (7, 4)

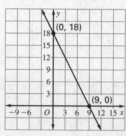

18. 0   **19.** Any horizontal line with an equation of the form $y = a$, $a \neq 0$.

Study Guide

1.
x-intercept $= -2$;
y-intercept $= 6$;

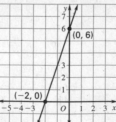

2.
x-intercept $= 9$;
y-intercept $= -15$;

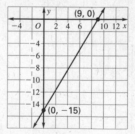

3. x-intercept $= 6$; y-intercept $= 4$;

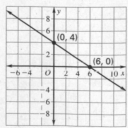

Lesson 8.3 *continued*

4. $20x + 10y = 100$;

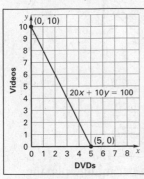

Sample answer: Three options are to buy 1 DVD and 8 videos, or 2 DVDs and 6 videos, or 3 DVDs and 4 videos.

Challenge Practice

1. x-intercept: -2;
  y-intercept: 4;

2. x-intercept: $\frac{5}{8}$;
  y-intercept: $\frac{5}{3}$;

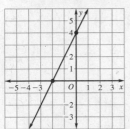

3. x-intercept: 2;
  y-intercept: $-\frac{15}{4}$;

4. x-intercept: 8;
  y-intercept: 4;

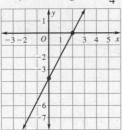

5. *Sample answer:*
  ; $(0, 2)$ and $(2, 0)$

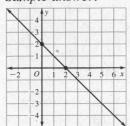

6. *Sample answer:*

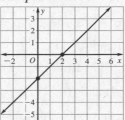
  ; $(0, -2)$ and $(2, 0)$

7. *Sample answer:* $y = 1 - x$

8. *Sample answer:* $y = x - 2$

9. The rectangle with the greatest whole number length has a length of 5 centimeters and a width of 1 centimeter. The rectangle with the greatest whole number width has a length of 1 centimeter and a width of 5 centimeters.

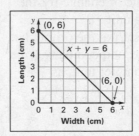

Lesson 8.4

Practice A

1. positive; $\frac{5}{4}$ **2.** 0 **3.** undefined

4. *Sample answer:* $(3, 12)$ and $(0, 6)$; 2

5. *Sample answer:* $(-1, 3)$ and $(1, 3)$; 0

6. *Sample answer:* $(0, -7)$ and $(4, -2)$; $\frac{5}{4}$

7. *Sample answer:* $(1, 1)$ and $(-2, 3)$; $-\frac{2}{3}$

8. *Sample answer:* $(5, 15)$ and $(10, 21)$; $\frac{6}{5}$

9. *Sample answer:* $(2, 3)$ and $(2, -3)$; undefined

10. $\frac{2}{3}$ **11.** 5 **12.** 0 **13.** -3 **14.** $-\frac{1}{2}$

15. undefined **16.** $\frac{3}{2}$ **17.** $\frac{3}{2}$ **18.** -4

19. $-\frac{23}{21}$ **20.** $-\frac{16}{19}$ **21.** $\frac{10}{9}$ **22.** $\frac{15}{8}$

23. a. 55 **b.** The slope represents the rate of change, or 55 miles per hour. **c.** *Sample answer:* It would start at the origin, but rise less steeply because the second car is traveling at a lower speed, and speed is indicated by the slope.

Lesson 8.4 *continued*

Answers

Practice B

1. undefined **2.** positive; 4 **3.** negative; $-\frac{7}{5}$

4. *Sample answer:* (1, 8) and (0, 11); -3

5. *Sample answer:* $(-1, -17)$ and $(1, -17)$; 0

6. *Sample answer:* $(8, -4)$ and $(0, -11)$; $\frac{7}{8}$

7. *Sample answer:* $(0, 7)$ and $(8, -2)$; $-\frac{9}{8}$

8. *Sample answer:* $(10, -1)$ and $(10, 1)$; undefined **9.** *Sample answer:* $(0, -21)$ and $(49, 0)$; $\frac{3}{7}$ **10.** 2 **11.** 1 **12.** -3 **13.** $\frac{2}{5}$

14. $\frac{1}{2}$ **15.** $\frac{13}{23}$ **16.** $-\frac{5}{16}$ **17.** 3 **18.** $-\frac{3}{2}$

19. $-\frac{5}{3}$ **20.** $\frac{4}{5}$ **21.** $\frac{3}{5}$ **22.** $\frac{5}{3}$ **23. a.** $\frac{7}{2}$

b. The slope represents the rate of change for the cost of the equipment, $3.50 a unit.
c. *Sample answer:* It would start at the origin, but rise less steeply because the second company is paying less for the fixed amount per unit, as indicated by the slope.

Practice C

1. *Sample answer:* $(-2, 3)$ and $(9, -1)$; $-\frac{4}{11}$

2. *Sample answer:* $(0, -14)$ and $(-3, -15)$; $\frac{1}{3}$

3. *Sample answer:* $(10, 0)$ and $(35, -9)$; $-\frac{9}{25}$

4. *Sample answer:* $(-4, -3)$ and $(3, -20)$; $-\frac{17}{7}$

5. *Sample answer:* $(-4, 1)$ and $(1, 4)$; $\frac{3}{5}$

6. *Sample answer:* $(-4, 11)$ and $(-1, 19)$; $\frac{8}{3}$

7. 5 **8.** -2 **9.** $\frac{5}{26}$ **10.** 12 **11.** $\frac{3}{4}$ **12.** $-\frac{1}{5}$

13. $-\frac{2}{3}$ **14.** $-\frac{4}{5}$ **15.** $-\frac{11}{4}$ **16.** $\frac{3}{31}$ **17.** -1

18. 1 **19.** $\frac{15}{26}$ **20. a.** Yes, a road sign is needed because the slope of the road is 8.4%.
b. 180 ft **21.** No; *Sample answer:* The second line could have a slope that is a negative number with a larger absolute value than the slope of the first line. Then the second line would be steeper.

Study Guide

1. $\frac{3}{2}$ **2.** $-\frac{4}{5}$; negative **3.** 0; zero

4. $\frac{5}{6}$; positive **5.** undefined

6.

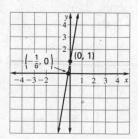

$m = 9$; You burn 9 Calories per minute playing basketball.

Challenge Practice

1. $-\frac{1}{3}$ **2.** 3 **3.** $-\frac{2}{5}$ **4.** $\frac{3}{5}$ **5.** $\frac{1}{2}$ **6.** $-\frac{3}{2}$

7. 1 **8.** -2 **9.** -7

10. *Sample answer:* **a.** $(-2, -4)$ **b.** $(-1, -5)$ **c.** $(-2, 0)$ **d.** $(-1, 1)$

Lesson 8.5

Activity Master

1. slope: -1; y-intercept: 4;

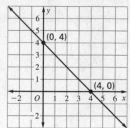

2. slope: $\frac{1}{3}$; y-intercept: -2;

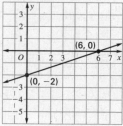

3. slope: 6; y-intercept: 1;

4. Yes; the slope is equal to the coefficient of x in the equation. Also, the y-intercept is equal to the constant term in the equation.

Lesson 8.5 *continued*

5. slope: -2;
y-intercept: 5;

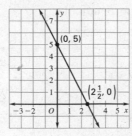

6. 2; no **7.** -5; no

8. $y = -2x + 5$; Yes. The slope is only equal to the coefficient of x in linear equations in function form. Similarly, the y-intercept is only equal to the constant term in linear equations in function form.

9. slope: 4; y-intercept: -3

Practice A

1. perpendicular **2.** slope: -5; y-intercept: 0

3. slope: 2; y-intercept: 1 **4.** slope: -4;
y-intercept: -2 **5.** slope: -1; y-intercept: 5

6. slope: $-\frac{2}{3}$; y-intercept: 2 **7.** slope: 2;
y-intercept: -4 **8.** C **9.** B **10.** A

11. slope: 3;   **12.** slope: $-\frac{1}{2}$;
y-intercept: 2;   y-intercept: 3;

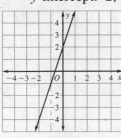   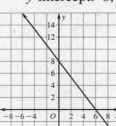

13. slope: 2;   **14.** slope: 2;
y-intercept: 12;   y-intercept: 6;

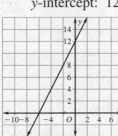   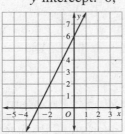

15. parallel line: $m = 6$; perpendicular line:
$m = -\frac{1}{6}$ **16.** parallel line: $m = \frac{3}{4}$;
perpendicular line: $m = -\frac{4}{3}$ **17.** parallel line:
$m = 1$; perpendicular line: $m = -1$

18. parallel line: $m = -3$; perpendicular line:
$m = \frac{1}{3}$ **19. a.** $y = -2x + 22$ **b.** 18 inches

Practice B

1. slope: $-\frac{1}{3}$; y-intercept: 6

2. slope: $\frac{3}{4}$; y-intercept: 0

3. slope: 4; y-intercept: -8

4. slope: 3; y-intercept: -12

5. slope: $-\frac{1}{3}$; y-intercept: 2

6. slope: $-\frac{3}{5}$; y-intercept: 3

7. B **8.** C **9.** A

10. slope: $\frac{5}{4}$;   **11.** slope: $\frac{3}{2}$;
y-intercept: 1;   y-intercept: 3;

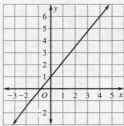

12. slope: $-\frac{4}{3}$;   **13.** slope: $\frac{1}{3}$;
y-intercept: 8;   y-intercept: -3;

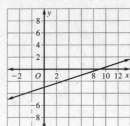

14. parallel line: $m = 12$;
perpendicular line: $m = -\frac{1}{12}$

15. parallel line: $m = \frac{6}{5}$;
perpendicular line: $m = -\frac{5}{6}$

16. parallel line: $m = 0$;
perpendicular line: $m = $ undefined

17. parallel line: $m = 1$;
perpendicular line: $m = -1$

18. parallel line: $m = -\frac{3}{8}$;
perpendicular line: $m = \frac{8}{3}$

19. parallel line: $m = -\frac{2}{3}$;
perpendicular line: $m = \frac{3}{2}$

20. a. $y = \frac{3}{4}x + 82$ **b.** 12:00 P.M.

Practice C

1. slope: 0;
y-intercept: 11;

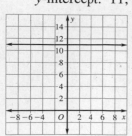

2. slope: undefined;
y-intercept: none;

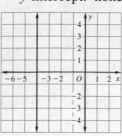

3. slope: −1;
y-intercept: −13;

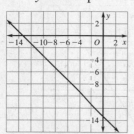

4. slope: $\frac{4}{7}$;
y-intercept: $\frac{4}{7}$;

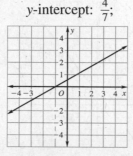

5. slope: 3;
y-intercept: −4;

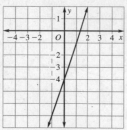

6. slope: $\frac{1}{5}$;
y-intercept: 3;

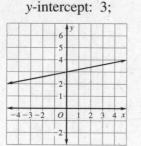

7. parallel line: $m = 0$;
perpendicular line: $m =$ undefined

8. parallel line: $m =$ undefined;
perpendicular line: $m = 0$

9. parallel line: $m = -\frac{5}{11}$;
perpendicular line: $m = \frac{11}{5}$

10. parallel line: $m = \frac{8}{13}$;
perpendicular line: $m = -\frac{13}{8}$

11. parallel line: $m = -\frac{2}{3}$;
perpendicular line: $m = \frac{3}{2}$

12. parallel line: $m = \frac{1}{17}$;
perpendicular line: $m = -17$

13. parallel **14.** neither **15.** perpendicular

16. perpendicular
17. a. $y = 5x + 40$

b.

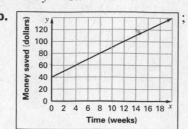

The slope represents the amount put in the jar each week. The y-intercept represents the initial amount in the jar. **c.** 18 weeks **d.** The graphs would be parallel and the y-intercept of the graph from eight weeks ago would be 0.

18. $y = \frac{3}{5}x + 4$ **19.** $y = \frac{3}{2}x - 4$

Study Guide

1. $m = -5; b = 4$;

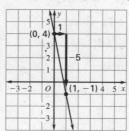

2. $m = 1; b = 1$;

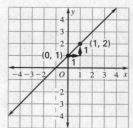

3. $m = -3; b = 3\frac{1}{3}$;

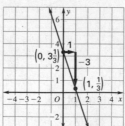

4. $m = 4; b = -2$;

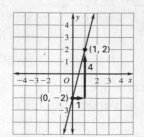

5. $y = 3x + 20$; You can play at most 6 games.
6. 2; $-\frac{1}{2}$ **7.** $-\frac{7}{11}; \frac{11}{7}$ **8.** $-\frac{5}{2}; \frac{2}{5}$ **9.** −1; 1

Real-World Problem Solving

1. x-intercept: 24; y-intercept: −6; Cara breaks even when she sells 24 cups of lemonade and Cara loses $6 when she sells no lemonade.

2. $m = 0.25$; Cara makes money at a rate of $.25 per cup.

Lesson 8.5 *continued*

3.

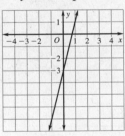

4. 20 cups: −$1; 30 cups: $1.50; 40 cups: $4

5. $1: 28 cups; $10: 64 cups; $100: 424 cups

Challenge Practice

1. slope: $\frac{4}{3}$;

y-intercept: $\frac{5}{6}$;

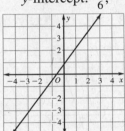

2. slope: 7;

y-intercept: −3;

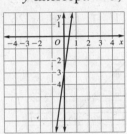

3. slope: 4;

y-intercept: −3;

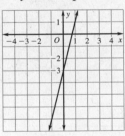

4.

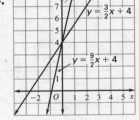

$y = \frac{3}{2}x + 4$

$y = \frac{9}{2}x + 4$

5. parallel line: $m = \frac{9}{4}$;

perpendicular line: $m = -\frac{4}{9}$

6. parallel line: $m = 2$;

perpendicular line: $m = -\frac{1}{2}$

7. parallel line: $m = \frac{10}{9}$;

perpendicular line: $m = -\frac{9}{10}$

8. parallel line: $m = -\frac{3}{4}$;

perpendicular line: $m = \frac{4}{3}$

9. parallel line: $m = -\frac{12}{7}$;

perpendicular line: $m = \frac{7}{12}$

Lesson 8.6

Practice A

1. sometimes   **2.** $y = x + 5$ **3.** $y = -2x + 8$

4. $y = 3x - 10$ **5.** $y = -x - 7$ **6.** $y = 3x - 1$

7. $y = -x - 3$ **8.** $y = \frac{1}{2}x + 1$ **9.** $y = -2x + 6$

10. $y = \frac{4}{5}x + 9$ **11.** $y = 4x + 10$

12. $y = -3x + 3$ **13.** $y = \frac{4}{3}x - 7$

14.

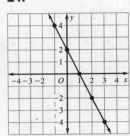

In a scatter plot, the data points lie in a nonvertical line, so the table represents a linear function; $y = -2x + 2$.

15. *Sample answer:*

a.

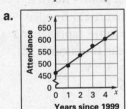

b. $y = 35x + 463$

c. 778 scientists

Practice B

1. $y = -3x - 2$ **2.** $y = 5x + 7$

3. $y = -\frac{3}{4}x + 3$ **4.** $y = \frac{5}{2}x - 6$

5. $y = \frac{1}{4}x + 2$ **6.** $y = -x - 1$ **7.** $y = \frac{3}{2}x - 1$

8. $y = -\frac{1}{3}x + 4$ **9.** $y = -4x + 5$

10. $y = -2$ **11.** $y = -\frac{2}{3}x + 9$ **12.** $y = \frac{3}{2}x - 2$

13.

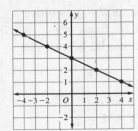

In a scatter plot, the data points lie in a nonvertical line, so the table represents a linear function; $y = -\frac{1}{2}x + 3$.

Lesson 8.6 *continued*

14. *Sample answer:*

a.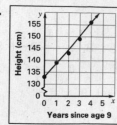

b. $y = \frac{23}{4}x + 133$

c. 179 cm

d. 127 cm

Practice C

1. $y = -\frac{5}{3}x - 5$ **2.** $y = \frac{8}{5}x - 12$

3. $y = -\frac{2}{7}x + 24$ **4.** $y = \frac{6}{7}x + 11$

5. $y = -\frac{8}{7}x - 16$ **6.** $y = -\frac{6}{7}x + 4$

7. $y = -\frac{3}{8}x - 7$

8.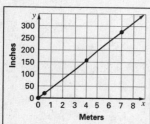

In a scatter plot, the data points lie in a nonvertical line, so the table represents a linear function; $y = 39.37x$.

9. Best-fitting lines may vary.

a.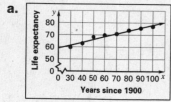

b. Using (30, 60) and (100, 77): $y = \frac{17}{70}x + 53$

c. about 2011

10. *Sample answer:* 45 pounds

11. No. *Sample answer:* The girl is likely to stop gaining weight at the same rate as she gets older, so the best fitting line for the data from age 2 through age 6 is not likely to model her weight accurately at the age of 25.

Study Guide

1. $y = 7x - 8$ **2.** $y = -\frac{5}{7}x - 4$

3. $y = -\frac{3}{8}x + 4$

4. *Sample answer:* $y = \frac{3}{5}x - 1$; $-6\frac{2}{5}$

Challenge Practice

1. $y = -6x + \frac{5}{2}$ **2.** $y = -\frac{5}{6}x - \frac{8}{3}$

3. $y = 8x - 2.1$ **4.** $y = 2x + \frac{3}{2}$

5. $y = -\frac{2}{3}x + 4$ **6.** $y = -\frac{8}{5}x - 2$

7. $y = \frac{3}{4}x - 3$ **8.** neither

9. $y = -x + 1$; First find the slope of the line. Then find the y-intercept by graphing the line.

Lesson 8.7

Practice A

1. $y = -x + 6$; $f(x) = -x + 6$

2. a horizontal line   **3.** -18 **4.** 2   **5.** 0

6. 1   **7.** 6   **8.** -26 **9.** B   **10.** C   **11.** A

12.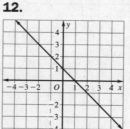

13.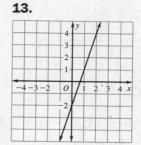

14.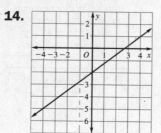

15. $f(x) = 2x - 3$ **16.** $f(x) = -\frac{1}{2}x + 5$

17. $g(x) = -\frac{3}{4}x - 2$ **18.** $h(x) = 3x + 28$

19. a. $d(t) = 60t$ **b.** $4\frac{2}{3}$ hours

Practice B

1. 21   **2.** 32   **3.** -3 **4.** 4   **5.** -18 **6.** 10

7.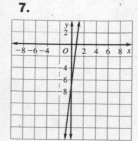

8.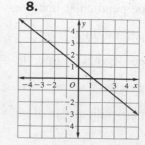

Lesson 8.7 continued

9.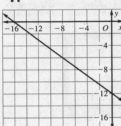

10. $f(x) = \frac{5}{3}x - 2$ **11.** $f(x) = -\frac{1}{6}x + 3$

12. $f(x) = -\frac{9}{4}x + 7$ **13.** $f(x) = \frac{5}{6}x + 40$

14. $f(x) = \frac{4}{7}x + 12$ **15.** $d(x) = \frac{7}{13}x - 2$

16. $g(x) = 7x + 111$ **17.** $f(t) = 8500t + 20{,}000$

18. a. $f(x) = 6x + 4120$ **b.** $71\frac{2}{3}$ minutes

Practice C

1. 110 **2.** −77 **3.** 4 **4.** 3 **5.** −2 **6.** 134

7.

8.

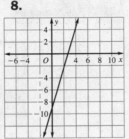

9.

10. $f(x) = -\frac{7}{9}x + 36$ **11.** $f(x) = \frac{25}{18}x + \frac{7}{2}$

12. $f(x) = 12x + 5$ **13.** $f(x) = \frac{5}{2}x + \frac{1}{2}$

14. *Sample answer:*
$f(x) = 3x + 15;\ g(x) = 4x + 15$

15. $a(x) = 3,\ b(x) = 4x - 13,\ c(x) = -\frac{4}{3}x + 3$

16. a.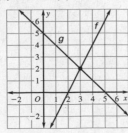

b. (3, 2) **c.** $f(x) = 2x - 4$ $g(x) = -x + 5$

$2 \stackrel{?}{=} 2(3) - 4$ $2 \stackrel{?}{=} -3 + 5$

$2 = 2\ ✓$ $2 = 2\ ✓$

17. $f(x) = -\frac{4}{3}x + 3;$ *Sample answer:* $(3, -1)$

Study Guide

1. 3 **2.** −17

3.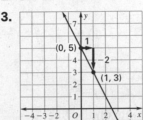

4. $f(x) = -x + 10$ **5.** $g(x) = -\frac{2}{3}x - 8$

6. $c(g) = 1.5g + 1;$ 8 gallons

Challenge Practice

1. 16 **2.** 4 **3.** $-\frac{17}{3}$ **4.** $-\frac{47}{12}$

5. $f(x) = 3.5x - 4.8$ **6.** $g(x) = -\frac{5}{4}x + \frac{2}{3}$

7. $h(x) = \frac{37}{60}x + \frac{3}{4}$ **8.** $g(x) = \frac{2}{3}x + \frac{10}{3}$

9. $P(w) = 8w + 4;$ When $w = 4.5,$ $P(w) = 40.$

Lesson 8.8

Practice A

1. A system of two linear equations can have exactly one solution, no solutions, or infinitely many solutions.

2. infinitely many **3.** yes **4.** no **5.** no

6. (2, 3) **7.** (2, 1) **8.** (−4, 4) **9.** (0, 0)

10. (1, −1) **11.** (1, 3) **12.** no solution

13. all points on the line $y = -3x + 1$

14. (2, 0) **15.** (3, 5) **16.** all points on the line $y = \frac{1}{2}x + 2$ **17.** no solution

18. a. $y = 12x$
     $y = 2x + 120;$ The solution of the system is (12, 144), so the cost of $144 for 12 visits is the same for each option. **b.** Option B is less expensive for 13 or more visits.

Practice B

1. yes **2.** no **3.** no **4.** (3, 4)

5. all points on the line $y = -3x - 7$ **6.** $(0, 3)$

7. $(5, -2)$ **8.** all points on the line $y = \frac{3}{5}x + 1$

9. $(-3, -4)$ **10.** no solution **11.** no solution

12. $(2, -1)$

13. a. $y = 8x + 240$ **b.** 45 family members
$y = 600$

c. less than 45 family members attend; more than 45 family members attend

14. $5\frac{1}{2}$ hours **15.** $(1, -3), (3, 1), (-1, 5)$

Practice C

1. yes **2.** yes **3.** no **4.** $(8, -5)$

5. no solution **6.** all points on the line
$y = 6x + 105$ **7.** $(5, 13)$ **8.** $(25, 30)$

9. $(-5, -7)$ **10.** no solution **11.** all points
on the line $y = -\frac{4}{5}x + 16$ **12.** $(-6, -14)$

13. infinitely many values of m and infinitely many values of b **14.** one value of m and infinitely many values of b **15.** one value of m and one value of b

16. a. $x + \quad y = 560$
$0.71x + 1.02y = 496.80$

b. aluminum: 240 lb; copper: 320 lb

c. $484.40

Study Guide

1. Infinitely many solutions; any point on the line $y = 7x + 9$ represents a solution.

2. $(-1, -10)$ **3.** no solution **4.** Festival A is a better deal when riding more than 4 rides.

Challenge Practice

1. $(2, -2)$ **2.** $(3, 2)$ **3.** $(-1, -3)$ **4.** $(1, 2)$

5. Infinitely many solutions; any point on the line $y = -\frac{4}{3}x + 8$ represents a solution.

6. no solution **7.** *Sample answer:* **a.** $y = 3$
b. $y = 2x$ **c.** $2y = 4x - 6$

8. length: 21 units; width: 7 units

Lesson 8.9

Technology Activity

1. 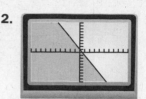 **2.**

3. 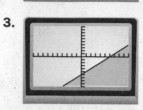 **4.**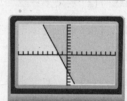

5. 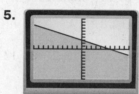 **6.**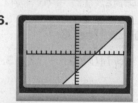

7. $1\frac{1}{2}x + \frac{3}{4}y \le 12$;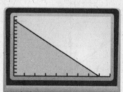

Practice A

1. *Sample answer:* Because the inequality is $<$, the boundary line should be dashed, not solid.

2. *Sample answer:* The wrong half-plane is shaded. The half-plane to the left and above the boundary line should be shaded.

3. no **4.** yes **5.** yes **6.** no

7. 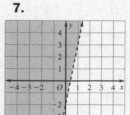 **8.**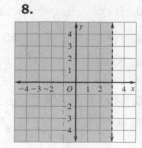

Lesson 8.9 *continued*

9.

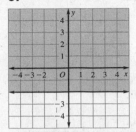

10.

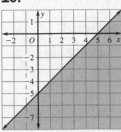

b.

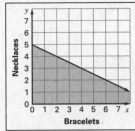

c. *Sample answer:* 0 bracelets and 5 necklaces, 2 bracelets and 4 necklaces, 8 bracelets and 1 necklace

20. Let x be the number of hours you read and y be the number of hours you write. Then $x + y \le 15$.

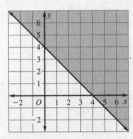

21. $x > -4$ **22.** $y \ge 1$ **23.** $y < x$

11.

12.

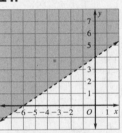

Practice B

1. *Sample answer:* The wrong boundary line was used. The boundary line should be the graph of $y = -\frac{5}{6}x + 3$.

2. *Sample answer:* Because the inequality is >, the boundary line should be dashed, not solid.

3. yes **4.** yes **5.** yes **6.** no

13.

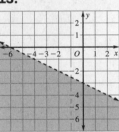

14.

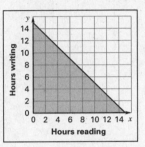

7.

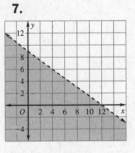

8.

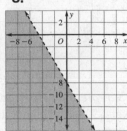

15.

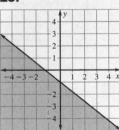

16.

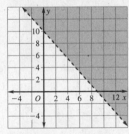

9.

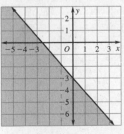

10.

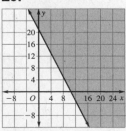

17.

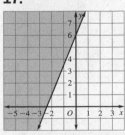

18.

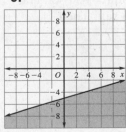

19. a. Let x be the number of bracelets you can make and y be the number of necklaces you can make. Then $x + 2y \le 10$.

Lesson 8.9 *continued*

11.

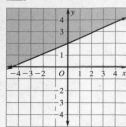

12.

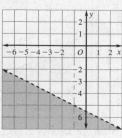

13.

14.

15.

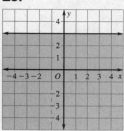

16. a. Let x be the number of adult tickets you buy and y be the number of child tickets you buy. Then $3x + 1.5y \leq 42$.

b.

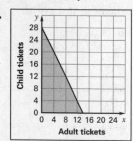

c. *Sample answer:* 1 adult ticket and 26 child tickets, 2 adult tickets and 24 child tickets, 5 adult tickets and 18 child tickets

17. Let x be the number of T-shirts you sell and y be the number of buttons you sell. Then $12x + 4y \geq 136$.

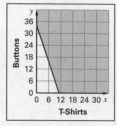

18. $x \geq -3$ **19.** $y < -2$ **20.** $y < x + 2$

Practice C

1. yes **2.** yes **3.** no **4.** yes **5.** no

6. yes

7.

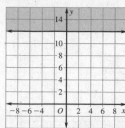

8.

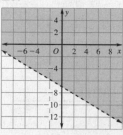

9.

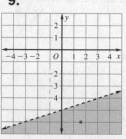

10.

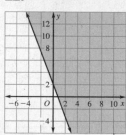

11.

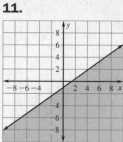

12.

13.

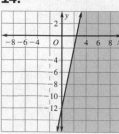

14.

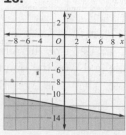

15.

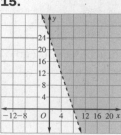

16.

Lesson 8.9 *continued*

17.

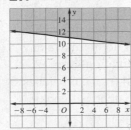

18.

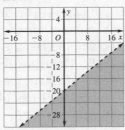

19. Let *x* be the number of hours you mow lawns and *y* be the number of hours you work at the grocery store. Then $5x + 6y \geq 82$.

Sample answer: 2 hours mowing lawns, 12 hours working at the grocery store, 8 hours mowing lawns, 7 hours working at the grocery store, 11 hours mowing lawns, 4.5 hours working at the grocery store.

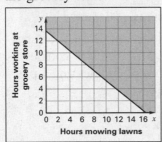

20. $y < -x$ **21.** $y \leq -3$ **22.** $y \geq x - 1$

23. a. no   **b.** yes   **c.** yes

24.

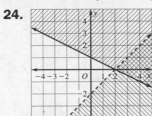

25. *Sample answer:* It is the intersection of two half-planes, and consists of all points that are below and to the right of the line $y = x - 2$ *and* that are also on or above and to the right of the line $y = -\frac{1}{2}x + 1$.

Study Guide

1. yes   **2.** no   **3.** no   **4.** yes

5.

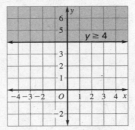

6.

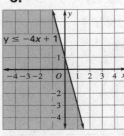

7.

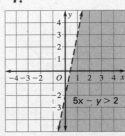

8.

Real-World Problem Solving

1. $y \leq 108 - 4x$

2.

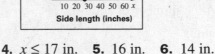

3. $y \leq 28$ in.

4. $x \leq 17$ in.   **5.** 16 in.   **6.** 14 in.

Challenge Practice

1.

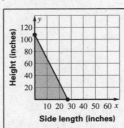

2.

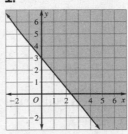

3.

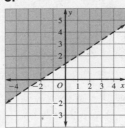

Lesson 8.9 *continued*

4. $y \geq 2x - 3$ **5.** $y < \frac{3}{4}x + 2$ **6.** $y \leq 6x - 4$
7. *Sample answer:* $y \geq 6 - x$;

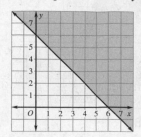

8.

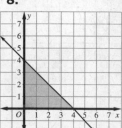

8 square units

9.

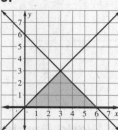

9 square units

10.

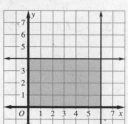

24 square units

Review and Projects

Chapter Review Games and Activities

Check students' work.

Real-Life Project

1. 5 cups **2.** $\frac{3}{4}x + \frac{1}{2}y \leq 5$

3.

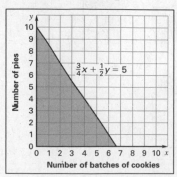

; Test points will vary.

4. The *y*-intercept represents the number of pies you can make if you do not make any cookies. The *x*-intercept represents the number of batches of cookies you can make if you do not make any pies. **5.** Three possible combinations are: (2, 7), 2 batches of cookies and 7 pies; (6, 1), 6 batches of cookies and 1 pie; (4, 4), 4 batches of cookies and 4 pies

6. $\frac{3}{4}x + \frac{1}{2}y \leq \frac{7}{2}$;

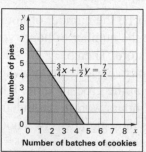

Test points will vary.

Three possible combinations are: (2, 4), 2 batches of cookies and 4 pies; (1, 5), 1 batch of cookies and 5 pies; (4, 1), 4 batches of cookies and 1 pie

7. Check student's work.

Cooperative Project

1–4. Check students' work.

Independent Extra Credit Project

1. Yes; every input is paired with exactly one output.

2.

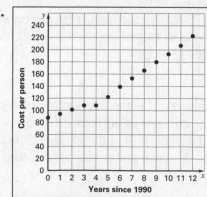

No, the points do not lie on a line.

3–8. Sample answers are given using $y = 11x + 78$.

3. $y = 11x + 78$

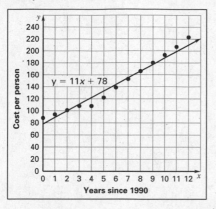

4.

Years since 1990, x	0	1	2	3	4
Cost per person, y	$78	$89	$100	$111	$122

Years since 1990, x	5	6	7	8
Cost per person, y	$133	$144	$155	$166

Years since 1990, x	9	10	11	12
Cost per person, y	$177	$188	$199	$210

5. $243   **6.** 2007   **7.** 2010

8. *Sample answer:* The amount spent per person for television costs is increasing each year with the biggest increases in recent years.

9. Check student's work.

Cumulative Practice

1. 14   **2.** 6   **3.** 3   **4.** 84   **5.** 28 goals

6. $x < -3$ **7.** $x \geq 8$ **8.** $x > 2$ **9.** $x \leq 60$

10. $\frac{3}{10}$ **11.** $\frac{4}{9}$ **12.** $\frac{3}{5}$ **13.** $\frac{14}{17}$ **14.** 8^{-1} or 2^{-3}

15. 121^{-1} or 11^{-2} **16.** $7p^{-3}$ **17.** $5x^2y^{-4}$

18. $\frac{14}{25}$ **19.** $1\frac{19}{50}$ **20.** $\frac{71}{1000}$ **21.** $-4\frac{21}{500}$

22. $\frac{2x}{3}$ **23.** $-\frac{12}{13y}$ **24.** $\frac{37k}{30}$ **25.** $\frac{9m}{20}$

26. 24   **27.** 35   **28.** 17   **29.** $\angle A$ and $\angle W$, $\angle B$ and $\angle X$, $\angle C$ and $\angle Y$, $\angle D$ and $\angle Z$, $\overline{AB}$ and $\overline{WX}$, $\overline{BC}$ and $\overline{XY}$, $\overline{CD}$ and $\overline{YZ}$, $\overline{DA}$ and $\overline{ZW}$

30. $\angle R$ and $\angle J$, $\angle S$ and $\angle K$, $\angle T$ and $\angle L$, $\overline{RS}$ and $\overline{JK}$, $\overline{ST}$ and $\overline{KL}$, $\overline{TR}$ and $\overline{LJ}$

31. 21   **32.** 40%   **33.** 50   **34.** 20%

35. $32.05   **36.** Domain: $-3, -1, 1, 3, 5$; Range: $-4, -2, 0, 2, 4$

37. Domain: $-0.2, 2.4, 5.2, 7.9$; Range: $-2.6, 0.1, 3.8, 4.7$

38. x-intercept: 2, y-intercept: 4;

39. x-intercept: 3, y-intercept: -2;

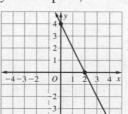
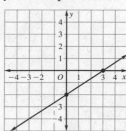

40. x-intercept: 3, y-intercept: -4;

41. x-intercept: 5, y-intercept: -5;

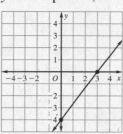
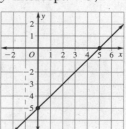

42. $y = 5x - 1$ **43.** $y = x + 2$ **44.** $y = \frac{2}{3}x - 4$